AF905154

ARCHAEOLINGUA

Edited by
ERZSÉBET JEREM and WOLFGANG MEID

Series Minor
33

Appropriate narratives
Archaeologists, publics and stories

Edited by

ELISABETH NIKLASSON and THOMAS MEIER

BUDAPEST 2013

Front Cover Illustration
Hidden narrative in the vicinity of the Berber village Imouzzer in Morocco
(Photo Elisabeth Niklasson)

Back Cover Illustration
Greek flag-design used during the Revolutionary War,
in a gun-slit of the *Kastro Favierou, Methana, Greece.*
(Photo Linos Papachristou)

ISBN 978-963-9911-47-5

HU-ISSN 1216-6847

2013

ARCHAEOLINGUA ALAPÍTVÁNY
H-1250 Budapest, Úri u. 49

Cover design by Erzsébet Jerem and Gergely Hős
Copyediting by Anna Freya Schneider, Elisabeth Niklasson and Thomas Meier
Desktop editing and layout by Rita Kovács

Printed in Hungary by Prime Rate Kft.

Table of Contents

Greece – for example ...

The dangers of narratives

Preface

This miscellany was triggered by a Round Table and collects papers from two sessions held at the EAA Annual meetings at The Hague (2010), Oslo (2011) and Helsinki (2012). We are deeply thankful to our co-organisers at these events, namely Carsten-Paludan-Müller, Petra Tillessen and Mark Pearce, for the great cooperation at these meetings and their kind support of this publication!

While we are used to books finishing and summing up a project, this book is something different: It tries to initiate discussion, to change perspectives and to conjure reflections on archaeology's relation and meaning to the wider public in a post-modern world. We neither carry with us the modern emphasis of educating the laymen, nor a patronizing perspective on the dull public which is constantly misunderstanding the important results of archaeologists. We also do not subscribe to the so called "anything goes" approach, a label which is really more of a straw man created by various opponents of post-modern approaches. Instead we were determined to take the publics appropriations of archaeology seriously – neither praising it to be the "real" archaeology, nor damning it to be false, neither regarding every appropriation to be legitimate, nor overlooking the dangers of some fringe appropriations.

We are aware that not all of our colleagues who have contributed to this volume share our perspective. We regard this to be one of the book's strengths (just as it was in the original sessions), not to unify different strands but to widen the focus and create a truly prismatic, multiperspectival view. Therefore, our special thanks go to the authors for joining us in this endeavour of diversity, for their valuable and thought-provoking papers – and not the least for their kind and considerate cooperation in responding to all of our editorial requests. We are sure that readers will enjoy their work.

Finally, our sincere "thank you" goes to Erzsébet Jerem for accepting this volume for the Series Minor of Archaeolingua. Moreover, we would like to gratefully acknowledge the dedication of our student assistant Anna Freya Schneider and her quick and concise editorial work, as well as the enthusiasm of Ágnes Anna Sebestyén, Rita Kovács and the Archaeolingua-team, who did final edits and – as usual – carefully supervised the production process.

Elisabeth Niklasson and Thomas Meier

Heidelberg and Stockholm, Midsummer's Day 2013

16th Annual Meeting of the European Association of Archaeologists, The Hague (The Netherlands), 1–5 September 2010

Abstract of the Round Table: "Archaeology and constructing (European) identity"

Organised by Carsten Paludan-Müller (NIKU Oslo, Norway),
Mark Pearce (University of Nottingham, United Kingdom),
Petra Tillessen (University of Heidelberg, Germany) and
Thomas Meier (University of Heidelberg, Germany)

What can archaeology offer to contemporary society? What could be our mission and relevance? What is the overarching theme of the development, role and potential of archaeology in Europe? What is the relation between archaeology's contribution to society and its explanations?

We believe that a discussion of such "meta-matters" of archaeology is now due.

As a first contribution, this Round table will focus on the question whether archaeology is a means to construct (European) identity.

Many archaeological narratives aim to explain the world as it is, with Europe as the outcome of a historical development, not to say a historical necessity. Thus archaeology provides stories of inclusion (Europeans) and exclusion (the others). To create such an European identity, archaeology will, in the opinion of some, have to highlight "European" as exclusive towards other cultures e.g. the Islamic world. In doing so, archaeology may emphasise that this is an old „natural" sharp division of cultures. But what makes Europe special might be its ability to integrate impulses and elements from other cultures: i.e. Europe (at its best) may not be defined in exclusiveness to other cultures, but rather as a particular dynamic way of dealing with/integrating diversity.

On the other hand, archaeology is a proper means to overcome nationalist(ic) histories, providing many examples that modern nation states are neither historical nor necessary geogrraphical entities. The question however, is whether it is inevitable that a European identity based on archaeology will be the same structurally as national identities based on archaeology, only on a different scale.

Practice shows that archaeology is a powerful and sucessful tool to create identities on a local and supra-national level. But is it so in an other than European context? Archaeology is a typically European method of telling stories about the past (e.g. our concept of time).

17th Annual Meeting of the European Association of Archaeologists, Oslo (Norway), 14–18 September 2011

Program of the Session: "Grand narratives in archaeology"

Organised by Thomas Meier (University of Heidelberg, Germany) and Carsten Paludan-Müller (NIKU Oslo, Norway)

The Geopolitical Perspective
Carsten Paludan-Müller, NIKU, The Norwegian Institute For Cultural Heritage Research, Norway

Who gets to tell the story? Ways of construction narration in Romanian archaeology from 19th to present
Monica Bira, National History Museum of Romania, Romania

Framing the past in the political space – On the "European Added Value" of narratives in the making
Elisabeth Niklasson, Stockholm University, Sweden

Apocalyptic narratives in archaeology – a mirror of public discourse
Thomas Meier, University of Heidelberg, Germany

Prehistoric wilderness – an ecocritical approach to Finnish archaeology
Tuija Kirkinen, University of Helsinki, Finland

Indus Valley Narratives: contesting the model of absoluteness through ancient gaming remains
Elke Rogersdotter, University of Gothenburg, Sweden

No narrative so grand
Geoff Carver & Matthias Lang, Universität Göttingen, Germany

18th Annual Meeting of the European Association of Archaeologists, Helsinki (Finland), 29 August – 1 September 2012

Program of the Session: "Public appropriations of archaeologists' narratives"

Organised by Thomas Meier (University of Heidelberg, Germany) and Elisabeth Niklasson (Stockholm University, Sweden)

Introduction
Elisabeth Niklasson, University of Stockholm, Sweden

Mediators

In search of the story
Gerhard Ermischer, University of Würzburg, Germany

Virtual and graphic reconstructions in public archaeology and the tension between science and public
Margaux De Pauw, Vrije Universiteit Brussel, Belgium

"Selecting what is important for the reader": about appropriations and transformations of archaeology in the mass media
Diane Scherzler, Südwestrundfunk, Germany

A report from a person who appropriates archaeologists' narratives for the public
Michael Cremo, Independent scholar, USA

Archaeology – appropriation and art

Bronze Age narratives – past and present
Joanna Sofaer, University of Southampton, United Kingdom & Marie Louise Stig Sørensen, University of Cambridge, United Kingdom

New glass houses for old stone balls: entangling the public with Neolithic carved stone balls
Mark A Hall, Perth Museum & Art Gallery, United Kingdom

Forward to the past: constructing heritage – Berlin Techno and the changing city
Luise Rellensmann, Technical University Cottbus, Germany & John Schofield, University of York, United Kingdom

Heritage in need of narratives?
Maria Persson & Anita Synnestvedt, University of Gothenburg, Sweden

Appropriations in sensitive political situations

Unshared knowledge and missed possibilities in Iranian archaeology
Mirette Modarress, University of Oulu, Finland

Public versus professional perceptions of an invisible heritage: a Greek case study
John Bintliff, Leiden University, The Netherlands

Forget the Parthenon! Alternative monumental roots to a "proper" Greek identity
Hamish Forbes, University of Nottingham, United Kingdom

Disarmed post-socialist archaeologies?
Social attitudes to interpreting the past
Eszter Bánffy, Hungarian Academy of Sciences, Hungary

Appropriation and rejection of stories

Archaeology and identity in a Bavarian village –
academic and local histories
Thomas Meier, University of Heidelberg, Germany

Neo-paganism, archaeological content and the belief in "Celts"
Jutta Leskovar, Oberösterreichisches Landesmuseum, Austria

Appropriated symbols – rejected narratives.
A discussion of public uses of rock art
Per Nilsson, Stockholm University, Sweden

Conversations *in situ*. Cultural heritage, tradition, mediation (poster)
Anna Gustavsson, Rio Kulturkooperativ; Sweden & Frederick Whitling, The Swedish Institute in Rome; Sweden

Appropriate narratives – an introduction

ELISABETH NIKLASSON – THOMAS MEIER

How this book came about – a teleological narrative

The story of this book started about four years ago in one of Europe's most beautiful landscape settings: It was in Riva del Garda, at what was probably the best organised EAA Meeting ever, that Carsten Paludan-Müller chaired a session on the implications of the Framework Convention on the Value of Cultural Heritage for Society (Faro 2005) for archaeology, and the meaning and place of our discipline in a wider social context. In the aftermath of an inspiring discussion, Mark Pearce and one of us (TM) independently contacted Carsten-Paludan-Müller with some loose ideas for another EAA session on this topic at the 2010 Annual meeting. It was in late 2009 that the three of us and Petra Tillessen joined efforts and learned of a recent debate within the EAA Board, stressing the necessity "to initiate a discussion of archaeology's potential for contributing to contemporary society. What could be our mission and relevance?"; more precisely the foci of the Board's reflections were on "'meta-matters' of archaeology [...] such as

- Archaeology's contribution to contemporary society
- Archaeology and 'grand explanations'
- Archaeology and identities".[1]

With the Board blowing wind into our sails we were discussing the potential of different formats, keeping in mind the wish for a strand of successive activities on several annual meetings in a row. As a first step we decided to start with a Round Table on "Archaeology and constructing (European) identity" held at the EAA-Annual meeting at The Hague in 2010. It was meant to trigger more initiatives in this field and resulted in two very well visited sessions at the following EAA-Annual meetings in Oslo (2011) and Helsinki (2012). The line of discussion in The Hague had tackled a wide variety of topics centering around questions of identity formation: the development of archaeology as a

[1] Both quotes from an Email by Carsten Paludan-Müller of 27th September 2009; forwarded 8th January 2010.

discipline in a specific European context, the archaeological quest for a cradle of Europeanness (contrasted to the Orient) in some distant past, the self-stylisation of archaeologists as secular priests of an arcane past, and the geopolitical implications of archaeology. As it was impossible to follow all these strands we quickly realised that it might be fruitful to concentrate on narratives of archaeologists. The Oslo session (organised by Carsten-Paludan-Müller and Thomas Meier) analysed "Grand narratives of archaeology", while the Helsinki session (organised by Thomas Meier and Elisabeth Niklasson) complementarily concentrated on "Public appropriations of archaeologists' narratives". To our happy surprise it was the Helsinki session which was most acknowledged in both the number of papers received and the number of visitors participating in the discussion – bestowing upon us the rare opportunity of a full day session.

While we felt it would be dissatisfactory to summarize a Round Table in a publication, this book is an attempt to wrap up the two sessions on archaeological narratives as they are closely related to each other. Clearly this publication is an enlarged selection: Some of the presentations dropped out for different reasons, due to lack of time or quality, or because they chose to publish in other journals. We have also included two more contributions in this volume which were accepted for Helsinki, but could not be delivered there for personal reasons (Ermischer, Katsarou-Tzeveleki). We are especially happy to make them public now! On the whole we have tried to collect a variety of authors from countries with very different traditions in archaeology-public-interactions. We have also tried to embrace authors with different levels of experience in different stages of their careers, as well as authors with differing social and professional backgrounds, including some with a professional background in fields other than archaeology. It was our aim to produce more than a mere miscellany of papers, to introduce a discussion and allow for contradictory views; if published in other places the provoking tension between some of the papers would not be able to develop their potentials for discussion.

Archaeologists, stories and publics

It has become popular over the last years to talk of archaeology as a 'craft' and of the properties of 'things', a discussion often referred to as a (material) 'turn' towards what archaeologists should really have been doing all along, and a way to 'reclaim' archaeology (see for ex. OLSEN 2010, 2012 etc.; GONZÁLES-RUIBAL *et al.* 2013). Although this may be regarded as a necessary reaction against widespread notions of culture as being mere text, or amounting to nothing more than a system of socially constructed symbols, these approaches seem to play down one of the main elements of the "craft" (or "art") of archaeology, the one of creating narratives. In order to convey their ideas archaeologists necessarily rely on language expressed in spoken or written form, or as visually articulated in sequences of images, and in order to be understood both inside and outside academia these have (consciously or unconsciously) employed narrative structures and meta-stories to connect things or events through coherent descriptions and explanations (see HOLTORF 2010; JOYCE 2002; PRAETZELLIS 1998; PLUCIENNIK 1999; 2010; VEIT 2010). Regarding academic reasoning we should be aware that even the notions of positivism and materiality are not based on objective ontologies (as they claim to be), but draw on hidden and socially constituted meta-stories as well (KARL 2010). Regarding both publics and politicians they explicitly ask for narratives: "Please tell the story!" as Nina Witoszek, Professor at the Center for Development and the Environment at the University of Oslo, put it in a nutshell at the opening ceremony of EAA's Oslo-meeting in 2011.

This plea is much more than a kind invitation to tell lovely stories about the past. It equally asks of archaeologists to live up to their promise and burdens them with taking responsibility for their stories and meta-stories (MEIER 2012). To accept this ethical perspective means to be aware of the narratological character of archaeologists' stories and to reflect on their meta-stories. Hayden White – following Paul Ricœur (1983/85) – has focused on the influence of contemporary meta-narratives[2] on the different emplotments employed in the historical

[2] In terminology we follow Cornelius Holtorf, who distinguishes meta-stories from meta-narratives – with some occasional overlap "The latter provide an overarching view of changes and continuities that are visible in the past, for example by portraying cultural evolution as progress or as decline, or by suggesting certain unchanging traits of human behaviour. The former, on the other hand, make archaeological practices, objects, and research results meaningful in an existential, social and political context in the present" (HOLTORF 2010: 383).

narratives produced during the 19th century (WHITE 1973; 1987; for a more thorough discussion cf. LAST 1995: 145–148; PLUCIENNIK 1999: 654). Although, if we look instead at the aims of meta-stories fostered by archaeological narratives we can see some other features:

1. The past as the same: There is nothing new under the sun and we are all part of a long and stable tradition! As this tradition is old, it is also proper and right and cannot or may not be changed. Due to its highly conservative character, this stance is hostile to change, it is "freezing" today's society in an everlasting chain of unaltered tradition.
2. The past as strange (LOWENTHAL 1985): The more distant it is, the stranger it gets and reversely, the smaller the temporal gap, the more familiar it appears. Such strangeness can be arranged in two ways:
   a. The past as bad old times: Past times were dark – as it must have been according to the logic of the enlightenment's imagination – the conditions of living were primitive, with people dwelling in caves or on trees with starvation omnipresent, daily life being dominated by mere struggle for survival. But in the course of history things got better. This narratological formula is clearly teleological, reaching its crescendo in contemporary civilisation. It is the modern narrative of progress, about the Western world as the climax of history. However, in contrast to type 1 narratives, this branch of type 2 generally allows for change, and the implicit climax of progress in today's society also raises anxiety about future changes, thereby simultaneously provoking a type of conservative self-satisfaction; we should be grateful to live here and now.
   b. The past as golden age: In the distant past man lived in accord with nature and the supernatural (the "noble savage" of Tacitus and David Thoreau). But in the course of history things got worse as man alienated from nature, from "the Gods" and finally from him/herself. Arranged in this way the past is a projection of the dreams of the present, and history is characterised by progressing decadence. Due to its implicit appeal it is either a narrative of frustration or a narrative of action, aiming to re-establish good old times, thereby also making it very conservative.

In consequence, archaeological narratives are not self-referential. They reaffirm the grand narratives of today's society and thus are fundamentally political: They explain the world in a specific way, (de)legitimising the social order (see Kirkinen this volume for an example of national narratives, and Niklasson this volume for a discussion on supranational narratives).

If archaeologists take these fundamental textual practises too lightly, considering them as mere vessels for the "true" description of "real" things, the transformations and appropriations of material culture in non-reflexive archaeological narratives – as well as the transformations and appropriations of archaeological narratives by other stakeholders and publics – may go unnoticed and escape responsible reflection. As a result, archaeologists could lose sight of an important factor in the production of knowledge as well as the values and responsibilities of archaeology. In the final discussion of the Helsinki session, in relation to an archaeological site being appropriated as an art exhibition hall, John Bintliff asked provocatively whether such a place really needs archaeologists' stories at all. While it is true that appropriations and transformations will occur with or without the involvement of archaeologists, we believe that one good way of engaging (or taking issue) with them is by creating our own powerful and convincing stories. Another one is to analyse the gaps and bridges between the narratives of archaeologists and those of publics. All contributions to this book represent different ways of doing so.

But what do we mean by 'publics'? The word is deliberately referred to in its plural form. This is done in order to avoid, to the extent it is possible, creating "them" as a clearly distinguishable and homogenous group that can be easily separated from archaeologists.

This distinction has a long history and many have argued that it is a built in function of the sciences to keep the public either away or at bay. Using the example of Socrates confronting Callicles in Plato's Gorgias, Bruno Latour connects this to "the fear of mob rule" and to the notion of the "mind-in-a-vat"; a mind that looks at the "real world" outside from inside a glass container, trying to reach certainty by observing an absolute reality separated from the way we "perceive" reality (LATOUR 1999). In the context of the current book, the fear of mob rule can be connected to archaeologists fear of losing interpretative precedence in a world where all interpretations count equally. Latour argues that this fear of losing touch with "reality" is a consequence of modernity which is based on the desire to silence and control the public, not a natural trait in human beings (LATOUR 1999: 10ff.). Furthermore, according to Richard Whitley, the more a field historically have sought recognition from peers and experts and cut themselves off from lay audiences, the more scientific they have been considered, the more centralised their knowledge production have been and the more resources they have gotten (WHITLEY 1984). Academic norms and values have thus directed research strategies and separated knowledge production from lay efforts so that truth becomes a monopoly of academics. Although archaeology as a discipline can hardly be called centralised, and have rarely enjoyed the type of funding reserved for more "scientific" pursuits, some still tend to think (despite the efforts of public archaeology) that what publics do with archaeological narratives is not something that should concern archaeologists.

So what about the name of the book? What do we mean by 'Appropriate narratives'?

Appropriate *Adjective:* suitable or proper in the circumstances.
Verb: take (something) for one's own use, typically without the owner's permission (OXFORD DICTIONARIES 2013).

In essence, there are no inappropriate narratives. Stories are told for different reasons at different times and places, and what makes them appropriate or inappropriate depends on the context as well as the 'yardstick' of the auditor. In the context of this book, the *adjective* form of this word is key. Our aim is to discuss, not discard, the narrators and narratives that are considered inappropriate in archaeological discourses, making the breakpoint – at which narratives are received, contested, or marginalised – into the object of this book.

While starting from the conception of narratives as in/appropriate, the discussion cannot manage without the *verb* form of the word, the act of "taking something for one's own use, typically without the owner's permission". However, this definition correlates poorly with the range of situations in which archaeological narratives are appropriated and with the way we wish to approach the topic. It makes little sense to talk of someone "owning" an archaeological interpretation, and they therefore cannot be "taken" for anyone's personal gain. A better definition in this case would be: to transform, make suitable, or engage a narrative – for a cause, an audience, or in a context other than the one originally intended. In this sense, to appropriate something becomes an open-ended practice and process in which archaeological narratives form but one part. But then, are there no "bad" narratives?

Taking a broad but critical perspective as our starting point, it is important to emphasise that this does not mean the equating or condoning of all types of narratives that draw on archaeology. To our mind, the yardstick also includes a mark dedicated to the consequences of narratives, contingent on ethical underpinnings and their ability to cause harm (with the Declaration on Human Rights as the lowest common denominator). In fact, such ethical considerations lie at the heart of recent discussions of appropriations in the fields of archaeology and cultural heritage (for ex. SCARRE – CONINGHAM 2012; YOUNG – BRUNK 2012[2009]). A focal point have long been the interaction between archaeologists and indigenous groups, especially in relation to colonialism, guided by the question of who owns the past (e.g. MCNIVEN – RUSSELL 2005; for a discussion on the US see WATKINS 2005; MCGHEE 2008). The question of why the appropriations or interpretations made by indigenous populations would be less harmful, more ethically correct or why their interpretations should somehow get special treatment has often been raised in these discussions (COLWELL-CHANTHAPHONH *et al.* 2010; TARLOW 2001; WILLIAMS 2009). This also actualises the question of scale; why do we so often favour the local over the regional, national or trans- and supranational?

Indeed, when it comes to countries within Europe like say Austria or Denmark, the special status of 'indigenous rights' becomes tricky, at least in the traditional meaning of genealogy, cultural attributes, continuity and self-identification. When the 'indigenous' population represents the majority, in/appropriate pasts can become very powerful means of personal and political expression, sometimes even dangerous (see BÁNFFY this volume). Archaeologist Sarah Tarlow has drawn attention to this situation, stating that "in much of Europe and Asia it is not the indigenous but the immigrant population who are socially and politically disadvantaged, and against whose interests the past may be used" (2001: 252, also in HOLTORF 2009: 674). This view is also supported by Cornelius Holtorf who argues that "the contemporary European perspective provides a strong argument for why indigenous perspectives must not be privileged" (2009: 673). Appearing contradictory at first glance, two years later he argues that the attention paid to minority groups has led to the interests of the majorities receiving less consideration, and that in the future archaeologists have to increase their understanding of general publics and address wider audiences (HOLTORF 2012). Thus, in a European context, the protection of minorities' rights or groups considered to be non-indigenous can be considered more ethically viable. At the same time we have to balance minorities' rights against the interests of the majority, partly due to their political and social advantage over the former, and partly due to the large and variable groups of people it represents (social class, subcultures etc.).

In this volume, the pivotal role of Hellenistic Greece and monuments like the Acropolis will be discussed as one such context in which the narratives and accounts supported by the majority population – self-identifying as indigenous – clashes with, or is appropriated by small local Greek societies or minority populations. These appropriations, by various groups at various levels of Greek society, are sometimes overshadowed by the ways in which this heritage has been adopted by other countries in Europe and the 'western world'. As three of the contributions will show, local and minority groups in Greece have chosen to deal with this hegemonic past in different ways, and strikingly, all of them emphasises the local level; be it specific ways of appropriation in the vicinity of a grand national monument (Katsarou), or local societies turning to local monuments rather than officially acknowledged ones, instilling in them both collective and personal meanings (Forbes, Bintliff). In connection to the latter, it is also observed that there are ways in which archaeologists can engage with such local narratives (Bintliff).

Aside from politically charged narratives and the importance of ethical considerations, this book also aims to highlight the role of other narrators engaging with archaeological accounts of the past. These narrators can be Neo-pagans (Leskovar), religiously inspired researchers (Cremo) or science journalists (Scherzler), all of which (although in different ways) tend to annoy many archaeologists. Not on ethical grounds per se, but because these narrators are not always interested in the truths of archaeologists but prefers to create their own truths about the past, or in the case of journalism, they have a very different set of goals than archaeologists. When they are not really considered harmful to others or to the monuments, it all seems to hinge on the accuracy of their beliefs and interpretations. It is the opinion of the initiators of this book, that while archaeologists admittedly do not need to treat such accounts of the past as equally valid, they do need to acknowledge their existence and speak to others with different understandings than their own, rather than shouting at them (for an example of archaeologists attitudes see Cremo this volume). So what kind of issues will be raised by the in/appropriate narratives to follow?

What's in the book

After a brief outline, hinting at some key themes within discussions on the appropriations of the past, we now turn to the contributions to this volume. This presentation is intended to highlight particular arguments and introduce the general topic of the papers, although as they are all very rich we will only refer to some of their dimensions. In fact, the papers could have been organised differently and the subdivisions of this book are merely a suggestion as to how they could relate to each other.

In the first section called the *The Nature of Narratives*, the central topic revolves around *how* stories are constructed, for which purposes and with the help of what enabling structures. In the first essay, **Gerhard Ermischer** takes us on a journey through a variety of stories from all over Europe, addressing their potentials and dangers as well as their make-up. Using vivid examples from his own interaction with publics in the German region of Spessart, he shows how "dull" glass shards can be brought to life in a story that connects to everything from globalisation to modern structures of transport. A warning hand is also raised against the ways in which narratives are sometimes used for propaganda, both in nationalistic settings and by supranational organisations like Council of Europe and the European Union. In the second contribution by **Elisabeth Niklasson**, the role played by the European Union in enabling narratives of archaeology and heritage is developed further. Does it matter from where we receive financial support, she asks? By looking into the criteria set for the European Commission funding programmes in culture, she examines what kind of narratives such criteria may encourage. Before that however, she reasons her way through a range of topics, such as perceptions of funding in archaeology, scholarly debates on Europe and different takes on archaeology and heritage in the European commission. In the third and last contribution to this section **Tuija Kirkinen** brings the analysis down to the level of the Nation state, drawing attention to the central idea of nature and wilderness to Finnish national identity and how this is expressed in archaeological narratives. Using an ecocritical approach she takes a close look at cultural constructions of nature in Finnish prehistory books and archaeological papers, finding that these narratives have been influenced by a range of transnational discourses.

In the *The stakeholders of narratives* we turn away from the creation of narratives and towards the creators, shedding light on different viewpoints and appropriations of archaeological narratives outside (or on the margins) of academia and professional archaeology. The first contribution by **Michael A. Cremo** is important for several reasons. Firstly because it is the only paper in this volume written by a person who has no formal education in archaeology, and secondly, because it is a reflexive account of what it can be like to handle both dismissive archaeologists and the transformation of their narratives. He provides us with a window through which we can look out at archaeologists and their narratives from the perspective of an "outsider". Of someone who is guided by a religiously grounded belief in extreme human antiquity and is often referred to as a 'pseudoscientist'. It is his firm belief that both he and the archaeologists he interacts with are better off for it. In the second paper **Diane Scherzler**, archaeologist turned science journalist – with a unique insight into two worlds – scrutinises the relationship between journalists and archaeologists. She pinpoints many of the misunderstandings occurring on both sides and opposes the perception of journalists as 'mediators' or 'appropriators' of archaeological stories or facts. Their job is to create interesting stories for publics she states, not to report the results of scientists, and it would make things easier if both sides better understood the premises of each other's work. Following Scherzler's discussion on the parallel views of archaeologists and journalists, **Thomas Meier**'s paper examines how academic and local histories can sometimes run in parallel, and how they can be preoccupied with completely different events in history and features in the landscape. He demonstrates, just like Scherzler and Cremo, that a division between publics, academics and mediators cannot be sustained, especially when looking at how narratives are used in constructing group identities at village level. The analysis and literary portrait of homeland researcher Karl Braßler provides a rich example of how one person can fill all three parts and more. The paper concluding the section is written by **Jutta Leskovar**. It addresses in depth how archaeological narratives are used in Neopagan literature, to support their belief in 'the Celts'. Interestingly she finds that archaeologists descriptions of prehistoric religion is not always that different from those of Neo-pagans, and that scholars therefore should be more explicit when they use such concepts. Similarly to the stories told by Cremo she describes the relationship between Neopagans and academic scholars as "messed up", and while she promotes dialogue between the two she thinks that archaeologists ought to focus on making their own stories heard in wider circles.

In Greece, for example ... the focus is, perhaps to no one's surprise, placed on appropriations and transformations of archaeological sites and narratives in the specific political and social milieu of Greece. At the time of writing (July

2013), the Greek nation is facing an enormously difficult situation which has led (among far more serious consequences) to the cultural heritage being increasingly activated in political discourses, and increasingly threatened by decay. The three papers that make up this section provide us with some insights into why this heritage is so central to Greek identities while at the same time being incorporated, contested or appropriated in different ways by local societies and minority groups. It is about inclusion and exclusion and about whose narratives count. The first thought provoking paper, written by **Stella Katsarou-Tzeveleki**, presents a project called the 'Young Neighbours of the Acropolis'. In this project the pupils of the 70th Primary School of Athens, located right next to the Acropolis hill, were encouraged to engage with the political campaign of reclaiming the Parthenon marbles. Rather than just presenting the outcome of this project, namely the intimate biographical images and texts created by the children to show their local attachment to the monument, she also discusses how such a project was made possible in the first place and on what assumptions it rests. While this project never really challenged the hegemonic view of the national Greek past (instead it might have reinforced it), it did show how close and personal the connection with the monuments can become, widely divergent from their otherwise 'universal values'. As if speaking directly to Katsarou-Tzeveleki, **John Bintliff** engages with the appropriations and transformations in another local Greek setting. However, he flips the coin and looks at the aftermath of the dominating role of Athens and the Classical past in the National Historical canon – at how rural communities have in many cases found themselves excluded from it and instead found their own ways of creating a meaningful past. Relating to notions of time and the role of memory, he provides us with three examples of how local monuments have been created or used to write local histories. More importantly, he also shows us how archaeologists and academic narratives can take part and become valuable sources of information in such ventures. In the final paper, **Hamish Forbes** calls attention to yet another form of local appropriation. Like Bintliff, Forbes is concerned with alternative narratives and monumentality, created by local societies to render their past meaningful in an environment where the authorised Classical past is the dominant Other. He shows how the Methanites, part of the Arvanitic ethnic minority which settled in Greece long before the independence, contests the Greek nation's categorisation of them as inferior citizens by turning to a monument from a different period, thereby allowing them to identify as "proper" Greek citizens who have fought for the nation.

In *The dangers of narratives,* **Eszter Bánffy**'s important and provocative report from Hungary stands on its own. It clearly links to the section on *The Nature of Narratives* in regards to issues like propaganda, political economy of archaeology and the culturally constructed narratives of the nation. It also relates to the second block through its focus on the intentions of stakeholders outside academia. Finally it connects to the third block as it addresses the consequences of wilfully distorted narratives in terms of who is included and excluded. In this case, such narratives do not only affect minorities, immigrants and other targeted groups, they also effectively shut the door for archaeologists' interpretations, attacking their viewpoints as well as their person and position. While representing a specific situation in a specific country, it is not without comparison. Far right wing groups, promoting views of a "racially homogenous" and pristine national past, are on the rise everywhere from Sweden to Greece and beyond. As the final contribution to this volume, it also intentionally leaves us with several open ended questions. Realistically, how and to what extent can archaeologists engage with ethically inappropriately appropriated narratives like these ones? One way is to continue to incisively and relentlessly ask the hard questions. Another, as has been mentioned earlier, is to write our own convincing narratives and try to reach out beyond our immediate circles. A third way, and the perhaps most important prerequisite for dialogue or at least understanding (not always condoning), is to beware of the multitude of appropriations within every arbitrarily categorised group. As we have seen, all Neo-pagan uses of archaeological narratives cannot be bundled together and what we have called 'stakeholders' actually represents a diverse group which can hardly be referred to based on their relation to archaeology. We have seen how differently the authorised Greek heritage is dealt with in different places and social and cultural settings and how it is often rejected. In the end, while highlighting the variability and dangers of appropriations of archaeological monuments and narratives may not have any direct impact on anyone but ourselves, we see this book as an important contribution to larger discussions on archaeology's role in society, encouraging self reflection in concert with outward engagement.

References

Colwell-Chanthaphonh, Chip – T. J. Ferguson – Dorothy Lippert – Randall H. McGuire – George P. Nicholas – Joe E. Watkins – Larry J. Zimmerman (2010), The premise and promise of indigenous archaeology. *American Antiquity* 75.2, 228–238.

Gonzáles-Ruibal, Alfredo (ed.) (2013), *Reclaiming archaeology: beyond the tropes of modernity*. London.

Holtorf, Cornelius (2009), A European perspective on indigenous and immigrant archaeologies. *World Archaeology* 41.4, 672–681.

Holtorf, Cornelius (2010), Meta-stories of archaeology. *World Archaeology* 42.3, 381–393.

Holtorf, Cornelius (2012), The past people want: heritage for the majority? In: Geoffrey Scarre – Robin Coningham (eds), *Appropriating the past: philosophical perspectives on the practice of archaeology*. Cambridge – New York – Melbourne – Madrid – Cape Town – Singapore – São Paulo – Delhi – Mexico City.

Joyce, Rosemary (2002), *The languages of archaeology. Dialogue, narrative, and writing*. Oxford – Malden.

Karl, Raimund (2010), *Macht und Ohnmacht des positivistischen Denkens. Der Positivismus in der deutschsprachigen Ur- und Frühgeschichte unter besonderer Berücksichtigung des Instituts für Ur- und Frühgeschichte der Universität Wien*. Beiträge zur Ur- und Frühgeschichte Mitteleuropas 58. Langenweisbach.

Last, Jonathan (1995), The nature of history. In: Ian Hodder – Michael Shanks – Alexandra Alexandri – Victor Buchli – John Carman – Jonathan Last – Gavin Lucas (eds), *Interpreting archaeology*. London – New York. 141–157.

Latour, Bruno (1999), *Pandora's hope: essays on the reality of science studies*. Cambridge.

Lowenthal, David (1985), *The past is a foreign country*. Cambridge – New York – Port Chester – Melbourne – Sydney.

McGhee, Robert (2008), Aboriginalism and the problems of indigenous archaeology. *American Antiquity* 73.4, 579–597.

McNiven, Ian J. – Lynette Russell (2005), *Appropriated pasts: indigenous peoples and the colonial culture of archaeology*. New York.

MEIER, THOMAS (2012), *Der Archäologe als Wissenschaftler und Zeitgenosse.* Darmstadt – Mainz.

OLSEN, BJØRNAR (2010), *In defense of things. Archaeology and the ontology of objects.* Lanham.

OLSEN, BJØRNAR (2012), After interpretation: remembering archaeology. *Current Swedish Archaeology* 20, 11–34.

Oxford Dictionary (2013), s.v. appropriate. http://oxforddictionaries.com/definition/english/appropriate [09.07.2013].

PLUCIENNIK, MARK (1999), Archaeological narratives and other ways of telling. *Current Anthropology* 40.5, 653–678.

PLUCIENNIK, MARK (2010), Is narrative necessary? *Ethnographisch-Archäologische Zeitschrift* 51, 48–63.

PRAETZELLIS, ADRIAN – MARY PRAETZELLIS (eds) (1998), Archaeologists as storytellers. Special issue of *Historical Archaeology* 32.1, 1–96.

RICOEUR, PAUL (1983/85), *Temps et récit*. Paris.

SCARRE, GEOFFREY – ROBIN CONINGHAM (eds) (2012), *Appropriating the past: philosophical perspectives on the practice of archaeology.* Cambridge.

YOUNG, JAMES O. – CONRAD G. BRUNK (2012 [2009]), *The ethics of cultural appropriation*. Chichester – Malden – Oxford.

WATKINS, JOE (2005), Through wary eyes: indigenous perspectives on archaeology. *Annual Review of Anthropology* 34.1, 429–449.

TARLOW, SARAH (2001) Decoding ethics. *Public Archaeology* 1.4, 245–259.

VEIT, ULRICH (2010), Zur Geschichte und Theorie des Erzählens in der Archäologie: Eine Problemskizze. *Ethnographisch-Archäologische Zeitschrift* 51.1/2, 10–29.

WHITE, HAYDEN (1973), *Metahistory. The historical imagination in nineteenth-century Europe*. Baltimore.

WHITE, HAYDEN (1987), *The content of the form. Narrative discourse and historical representation*. Baltimore.

WHITLEY, RICHARD (1984), *The intellectual and social organization of the sciences*. Oxford –Toronto.

WILLIAMS, JONATHAN (2009), Indigenous voices, archaeology, and the issue of repatriation. In: Chris Gosden – Barry Cunliffe – Rosemary A. Joyce (eds), *The Oxford handbook of archaeology*. Oxford.

A visit to the Motel of the Mysteries: Stories and storytelling in archaeology

GERHARD ERMISCHER

Abstract:
Telling stories is a very human activity – it is of great social importance and a driving force of human (cultural) evolution. Therefore it cannot come as a surprise that storytelling also lies at the heart of history. It is a powerful tool for historians and archaeologists to communicate their results, hypothesis and models to the public. But it is not only a prerogative of specialists or experts. People tell their own stories about historic and archaeological monuments all the time, having their own stories in mind about features and sites. These stories need to be taken seriously by experts and should be considered as part of doing archaeology, as part of the history of a monument or the monument itself. Since stories are powerful and can be very convincing, it is also necessary for the historian or archaeologist to remain aware of the fact that stories are constructs. Both in the sense that they themselves construct stories and use them for their own means, but also that others – who are of course at least as good as us (or even better) at constructing convincing stories – do so for their own purposes. This becomes especially visible (and dangerous) within national identity building processes, which often leads directly to the creation of nationalistic stories. Historians and archaeologists can easily be caught up in this process, either willingly because of the rewards offered, unwillingly through coercion, or due to plain negligence of their ethical responsibilities. This essay aims to show how powerful stories can be, how they can be used to communicate and why they should be respected, and in which contexts academics need to be especially careful.

> "Beware Great A'Tuin, the space turtle gliding through the universe. On her shell stand four elephants who carry on their shoulders the Discworld© – a world full of magic and fantasy" (from the lead of the film adaptation of Terry Pratchett's *Going Postal* [2010]).

The novelist Terry Pratchett has created this fantastic World in more than thirty books so far, with side stories and mini-series playing out within the great saga. But one element is consistent and dominant throughout the books: the narrativum. The narrativum is the essence of the stories, the story line, the story behind the stories, the fundamental force that holds the universe together, so desperately thought by Goethe´s Faust. In the book series called *The Science of Discworld*, Terry Pratchett popularises modern science, telling it as if it was just another story set in his fantastic universe. One of the three books, *Darwin´s Watch*, is dedicated to evolution theory (PRATCHETT 2005). In this book he explains that storytelling is the essence of human evolution. The ability to invent stories, tell stories and to understand the true meaning of stories is the driving force of human development, allowing for the transfer knowledge from generation to generation.

In *Hogfather*, Terry Pratchett builds a whole novel around the power of storytelling (PRATCHETT 1997). Of course, sooner or later all these stories are about blood – but they also allow humans to create abstract concepts like justice, mercy or love. Stories are powerful, Pratchett writes, and when stacked together in great numbers in small spaces (libraries) they bend the time-space continuum turning it into the 'L-space' (L for Library). This L-space can then be navigated by seasoned librarians in order to get access to all the books which were, are, and will be (and maybe even might be). Stories do have power, and in a very entertaining way Terry Pratchett actually teaches us the basic truth about history: that sooner or later it is all about stories.

Stories and archaeology

As archaeology is a historic science, it deals with both great and small narratives. It is important to understand that stories not only capture the interest of people, but that there are many different stories told by many different people: scientists, researchers, academics and scholars all tell their own stories (often in a style illegible to anybody else but another academic). But so do "local" people such as visitors to museums and sites, interested laymen, professional writers, poets, novelists and authors of historic novels. Even artists tell their own story about historic events and archaeological sites, especially if one looks at the paintings done in the spirit of 18th and 19th century Romanticism, where lonely dolmens (*Fig. 1*) played as great a part as the Egyptian pyramids and the wonders of the antic world. Archaeological sites and objects was also a theme for the fantastic Vedutes of Piranesi, and for the historistic interpretations by Gustaf Klimt and his colleagues in the entrance hall of the Museum of Arts in Vienna (Kunsthistorisches Museum Wien).

A wonderful way to make space for archaeological narratives is through the concept of the communal dig. I first experienced it in Manchester. In the towns

Fig. 1. Johan Thomas Lundbye - Dolmen at Raklev 1839,
romatic painting of Dolmen as a dominating feature of the Danish landscape
(Photo Thorvaldsen Museum, Copenhagen / author).

of the Greater Manchester Area there are currently many economic and social problems, and here the communal dig became a strategy to pro-actively fight these problems. A potential archaeological object was chosen, a monument which was not listed but which held a specific meaning for the local people. It was a cotton mill from the Victorian era, the time of industrialisation and booming textile production in the area. Everybody in the community was invited to participate in the excavation on their own terms. You could come for a few hours in the afternoon, for a couple of days in a row, every Saturday, just once, or you could show up every day throughout the excavation period. School classes were invited to take part in, or to visit the dig. A big fare was organized; a party for the whole community with participatory activities like re-enactments and historic crafts. The visitors also took part in guided tours, not only led by the professional archaeologists who supervised the excavation, but by their peers. As a result, a lot of different stories were told, both the stories of the experts and the stories of the volunteers who would speak about their motivation, experiences, their own interpretations and ideas. People came to share with us what they knew about the old cotton mill, for example how their grandmother had worked there. The actors told the stories of their fictive personalities and the craftsmen of their tradition and experience. There was even a grid outlining the site in preparation for next year´s excavation and everybody could grip a diving rod and walk along the lines to see where it might dip to the ground. While people did this, students were marking the places on the grid in their laptop. They wanted to see if there were any apparent clusters based on the reactions of the diving rods, which should then be checked against the results of the upcoming excavation. The project made room for all these different narratives and encouraged many persons, who normally never would have been interested in archaeology, to engage in the history or the cultural heritage of their region.

History is a sequence of stories, and while stories are often told in order to explain events and developments of the past, they are also very much attached to specific sites or certain types of sites. The Scandinavian stories about trolls living inside pre-historic grave mounds, or the stories about the "beds of Giants" (*Hünenbetten*) in Northern Germany, are as much part of the reality of such sites as the stories of their role as 'hoards of ancient wisdom' are to the pyramids. Furthermore, stories are attached to vernacular structures, something which has rarely been drawn attention to in the world of academia. During a project on cultural landscapes our colleagues in Halland (Southern Sweden) documented the stories woven around a simple, circular depression in the forest. The local people were fascinated enough by this peculiar structure to want to know what it was all about, thus creating their own stories to explain it: a historic wolf pit, a crater of a bomb from the Second World War, a prehistoric tomb (CARLIE – ROSENGREN 2003: 79–81).

Ultimately, the stories told about archaeological sites, structures or objects – while often being considered "meaningless" by academics due to their un-scientific character – are part of peoples' identities, part of building identities, making them significant to those who tell the stories and those who listen to the stories. This layer of reality has to be recognised and taken seriously, and in archaeological documentation it should be recognised as part of the object itself.

Money makes the world go round – but so do stories. Would Schliemann ever have looked for Troy without Homers Iliad? What would be the significance of a certain dusty and stony hill in Turkey without the story of Achilles, Hector and Helen of Troy? It would still be interesting for archaeologists for sure – but would it also be interesting for the wider public? Without their interest, and consequently that of their government, would there be enough money to excavate and research the place at such a grand scale? Would it still invoke the fierce academic discussions, neigh quarrels, it has? Stories endow a place with meaning, whether it is the stories of the place itself or the story of its finding. Would the treasures of Tut-ench-Amun ever have made such a great impression on so many people without the romantic story of the search and recovery of the tomb by Howard Carter and Lord Carnarvon? Would it have inspired authors to write romantic novels and movie makers to create stories of mummies resurrected?

Fig. 2a. Sophia Schliemann, wife of Heinrich Schliemann, poses with the treasure of Troy or "Treasure of Priamus" in 1873.

Fig. 2b. David Macauley's interpretation of the theme in his "Motel of the Mysteries" (1979). The wife of the excavator of the motel poses with the finds from the sacred grove (e.g. the bathroom of the motel suite), which are interpreted as ritual jewelry.

There is even a narrative of archaeology as such, so strong that it could be used by David Macauley in his brilliant illustrated story *Motel of the Mysteries*, this hilarious parody of archaeology (MACAULEY 1979) (*Fig. 2a, b*).

Stories of shattered glass

Stories make sense of an object, putting it in a framework; a network of information and interpretation. For instance, a glass shard excavated at a site in Antwerp can be interpreted as part of a drinking vessel of a specific form, which in turn can be dated, and finally a chemical analysis may help to establish its place of origin. All this information can be catalogued, making it important only to other researchers whom it may help in the process of dating and cataloguing their own finds. But put into context and it can become the focus of a great narrative!

The narrative of the Spessart glass huts for example (*Fig. 3*). Glass production in the Spessart became an important economic venture in the 12th century. Located in a forested German upland region, which is mainly known for its forests and its stories of poverty and highway men, it formed the basis of an early industrialisation (compare e.g. KRIMM 1982). Chemical analysis by Karl-Hans Wedepohl could prove that the great stained glass windows in the cathedral of Cologne were made from Spessart glass (WEDEPOHL 2003: 133–135). In late medieval and early modern times most of the glass production – by then more glass vessels then window panes – was directly sold to the Netherlands, mainly to Antwerp which was a commercial node of great importance. Traders from Antwerp and bankers from Frankfurt were dominant figures when it came to financing glass production in the Spessart, an investment which paid off well for the investors. This established a strong economic link between the production area in the Spessart and the consumer area in the Netherlands (LOIBL 2001).

For a long time the main item of production was green glasses used for drinking wine. This changed dramatically in the 17th century when the wars between the Netherlands and France created a breach of well-established trade routes between the South of France and the Netherlands. Wine was one of the major import goods from Southern France. This was also the time of Mercantilism, an economic philosophy insisting strongly on the keeping of one's money in one's own realm: making sure that you export more than you import. So the Dutch government had two good reasons to cut down the import of French wine and foster the production of Dutch beer (a local wine production was not really an option), and taxes and tolls to help speed up the process. Within a very short period the drinking habits

Fig. 3. Guided tour on the cultural path Kleinkahl including the Eppstein glass huts, excavated in advance of the opening of the thematic trail. Guides in historic costumes act out short plays to introduce the audience to the story of glassmaking in the Spessart (Photo ASP / Gerrit Himmelsbach).

changed and the Netherlands became a beer drinking nation. Beer was mainly consumed in white or transparent glasses, while wine, which was not filtered and not as brilliantly coloured as it is today, looked better in a green glass. The brown beer of the time looked best in a transparent glass and the saying that we eat and drink with our eyes as much as with our mouth, is as old as it is true. As a result the glass production in the Spessart rapidly changed, from making bulks of green wine glasses to bulks of white beer glasses (KRIMM 1984: 175).

But this is not all there is to this story. The trade with bulk loads of glass also required a well-developed capacity for transport. This is where the Frammersbach teamsters come in. They most likely started as transporters of glass to Antwerp. Thanks to other historic events, like the Hussite wars which cut off old trade routes to the east, they changed into European lorry drivers from the 16th century onward. They transported goods for the great trading houses in Augsburg and Nuremberg – like the Fuggers – from Venice to Antwerp, or to the great fares in Frankfurt, Cologne and Leipzig. As they were also responsible for paying tolls and taxes and dealing with all the legal issues along the way – with so many borders and trade-privileges in different towns and cities – they were indeed more than just lorry drivers and resembled something akin to modern logistic companies. In Antwerp they had their own guild hall which still exists today, the famous "Hessen Huis" (MOSER 1990).

Knowing all this we can now bring the glass shards to life, we can tell their story and make them part of a great and fascinating narrative. They are at this moment interlinked with stories of early globalisation, trade, financial networks and surprisingly enough also with modern structures of transport and logistics. Famous names like the Fuggers come into play; connections to greater history become visible, to European power plays of dominance, to wars and well-known figures like King Louis XIV of France. This is when history becomes tangible and when stories allow us to understand the significance of seemingly modest and humble archaeological fragments. When the Archaeological Spessart-Project excavated the castle of Partenstein in the Spessart, many objects were found. Each of them can in itself be used to tell a story, but putting all of them together allow us to weave this excavation site into this great narrative. The many fragments of glass melting crucibles found in the castle ditch point towards the importance of the glass production as a source of income for the founders of the castle, the earls of Rieneck. Lead seals from cloth bales and sacks of spice link the castle to Venice and Antwerp, while stone ware from the Rhineland link to Cologne, and high quality brass objects link to Nuremberg. When mapping the places of origin of all these objects, they overlap exactly with the main transport routes of the Frammersbach teamsters – Frammerbach being the next village to Partenstein (ROSMANITZ 2008).

As the Spessart is a region with a very strong and clear cut image, these stories are of special importance to people in the area. It was the hunting resort of the archbishops of Mayence and there was not much development until quite recent times. The only sources of income were those generated by glass production and charcoal burning. The archaeological and historic research of the last 15 years, undertaken by the Archaeological Spessart-Project, allows us to draw quite a different picture: a highly exploited economic landscape with early industrialisation, lined with important transport routes connected directly to the hot spots of economic development in Europe, like Venice, Augsburg,

Nuremberg, Frankfurt, Antwerp or Amsterdam. These interconnections and the ensuing dependencies had an enormous impact on the development in the Spessart (including of course historic events which people here had no chance to influence). It is a story of early globalisation, financial streams and economic networks, differing dramatically from the old story of poverty and isolation on the fringes of the civilised world. These enriched stories are now being told along the cultural paths (thematic trails) developed by the Archaeological Spessart-Project together with the local inhabitants (ERMISCHER – HIMMELSBACH 2009).[1]

Of course, this old image is in itself a result of storytelling. The Spessart became the iconic landscape for isolated woodland, hunting and highway-men in German literature at an early stage. Already in the Nibelungen Epos the Spessart is mentioned as a hunting resort. In other famous works of German medieval literature the Spessart is used as an icon for the dark woodland (for a selection of quotations of medieval literature mentioning the Spessart see HARTMANN 1935: 3–5). In the most influential novel of German Baroque writing, the story of *Simplicius Simplicissimus* by Grimmelshausen – reflecting on the atrocities of the Thirty Years' War – the Spessart is not only referred to as the birthplace of Grimmelshausen himself, but depicted as a forested wasteland which can serve as a refuge for those weary of the world, populated only by a few charcoal burners and a recluse (VON GRIMMELSHAUSEN 1669). In reality however, things were quite different, especially since the highways through the Spessart were colonized by the armies of the Thirty Years' War, carrying destruction, plundering and deceases to its villages (KAMPFMANN – KRIMM 1988: 63–66). In the 16th century the famous cobbler and poet Hans Sachs in Nuremberg wrote a poem about a young farmer who became a highway man in the Spessart, and who was to be executed in Frankfurt (the full poem is quoted in HARTMANN 1935: 9–11). No less a celebrity than Götz von Berlichingen wrote about ambushes he had planned and undertaken in the Spessart, targeting wealthy merchants from Nuremberg as well as the archbishop of Mayence (VON BERLICHINGEN 1562, the ambushes in the Spessart quoted by HARTMANN 1935: 8–9). His autobiography was published in the 18th century and immortalised by Johann Wolfgang Goethe in one of his most popular dramas. And in the Romantic era, when Wilhelm Hauff wrote his novel 'The Spessart Inn' (*Das Wirtshaus im Spessart*, HAUFF 1827), the Spessart was finally branded as the hotbed of highway men and robbers. Again, this did

---

1 You can also visit them on the internet at: http://www.spessartprojekt.de/kulturwege/index.php (20.07.2013).

not align with the historic reality. The Head of Police, responsible for catching most of the highway men who flourished in the politically highly fragmented Germany around 1800 – suffering from the social unrest of the French Revolution and the Napoleonic Wars – wrote a book about the great gangs of highway men in the Rhine Area. When it became a bestseller he did what all bestselling authors love to do – he wrote a sequel, this time about the highway men in the Spessart and Main Area. Already in the pretext he stresses that this was only a minor stage within what was known as the 'great plague of robbers' but that, seeing how successful his first book on the topic had been, he had decided to follow it up with a smaller tome on this area as well (PFISTER 1812: III–IV).

Here we can see the great power of the stories. They can create images, brand mark whole regions and invent identities. Of course, there are physical things and historic statements helping to create and upkeep these stories, but from that point on they can become quite independent – and once they have become strong narratives they are difficult to overcome. For many people in the Spessart this previous image of was not a positive one, even when it could be used as a trade mark in tourism. Therefore the "new" stories created based on recent archaeological and historic research are welcomed by many people in this region. They help in building an identity marked by pride, rather than shame or disinterest, in an area which has often been sold under value (ERMISCHER – ROSMANITZ 2007). The power of stories and of these particular narratives naturally also comes with a great responsibility for all who use them. Storytelling helps to create interest and make people aware of historical events, granting significance to archaeological sites and objects. But there is always a potential for misuse, something which has to be kept in mind at all times and carefully avoided. Identity building can be a very positive aspect on local and regional level. Especially in landscapes with a low profile where it helps to raise awareness and give people a sense of pride for their own landscape and cultural heritage. Based on a proactive strategy this can lead to a sense of ownership and responsibility, turning local people from disinterested bystanders into informed consumers, and finally into active wardens of their landscape and cultural heritage (ROSMANITZ 2011). On a national level identity building is, more often than not, misused for nationalistic propaganda in which the narratives of historic or archaeological sites are used as powerful tools.

This can be seen for example in the recent conflict between Greece and The Former Yugoslavian Republic of Macedonia.

Stories of resurrected golden pasts

In the agonising process of the decline of former Yugoslavia, one of the emerging new states was Macedonia. The name and identity given to this former autonomous region within Yugoslavia was created by Tito in order to block possible claims by neighbouring Bulgaria. When Macedonia became independent under this modern name, it caused uproar in Greece. Greece claimed to be the home of the one and only true Macedonia, home of Alexander the Great. There was *Angst* that the newly founded Republic of Macedonia could raise claims on the Greek province of Macedonia, and thus the Greek government fought nails and teeth against the name of the new state. The compromise was the uncanny monster of a name under which Macedonia still has to present itself to the world: The Former Yugoslavian Republic of Macedonia. But it also led to a nationalistic revival of the cult around Alexander the Great.

When I first visited Vergina, antic capital of antic Macedonia, to see the royal tombs of the dynasty of Alexander, it was a lonely wasteland covered with shrubs. The finds of the royal tombs were on display in a museum in Thessaloniki, a dire affair of old show cases in which the precious objects were stacked on top of each other, covered with dust. The one and only warden was an elderly man who told stories about his grandfather fighting against the Turks to liberate Greece, complaining about a lack of willingness in the Greek youth to step up to the plate in a recent conflict with Turkey. He bemoaned the state of Greece, governed by decadent southerners in Athens and corrupt politicians who had not a single straight bone in their body, and he claimed that the true Greek were the Macedonians, the descendants of Alexander the Great. He referred to a cask found in the tomb of Philipp, the father of Alexander, which bore an eight-pointed star which he claimed to be the coat of arms of his own village, making them direct descendants of the great king.When I came back 15 years later, for a European Association of Archaeologists conference, the objects had been removed from the museum to the archaeological site of Vergina. The site had become a national shrine. Despite the fact that it had become a UNESCO World Heritage site, the actual tomb of Philipp had been converted into an underground museum, pitch black, with the modern show cases emerging as highlighted islands glittering with the gold and jewels of the royal grave goods. Taking pictures was strictly prohibited and the uniformed wardens acted as guards of the sacred trove, chastising visitors for raising their voices above the respectful whisper. The dominating feature above ground was a huge Greek national flag. Of course, the refurbishing of an original grave mound into a museum is definitely against all rules of UNESCO World Heritage Sites, and the same should go for transforming them into national shrines. But it was obvious that this installation was directed against the new Republic of Macedonia, stressing the Greek claim to a historic "truth" on which the Greek nation was founded. A complicated claim, as some Greek leaders looked back on antiquity, others on the Byzantine heritage, all while the western European romanticists and others who supported the Greek struggle for independence – out of romantic visions of ancient Sparta and Athens – completely ignored the realities of modern Greece. The latter representing a conflict of ideas and realities, mirrored in negative reports of the German Philhellenic volunteers during the Greek war of independence – as evident in relation to the disposition of the first king of an independent Greece, the Bavarian Otto I (QUACK-MANOUSSAKIS 2003).

On the other hand, the Republic of Macedonia was not based on any reference to ancient Macedonia or Alexander the Great, but when they were forced to defend their name and national identity against the Greek claims, the newly founded state started to pan for a national heritage in the disputed source. Macedonia soon started to build a national myth based on ancient Macedonia and Alexander the Great, renaming streets and places of Macedonian towns and villages after heroes of the Macedonian Antiquity. A first hand report on archaeology in Macedonia, given by a colleague at the annual conference of the German Society for Pre- and Proto-history (*Deutsche Gesellschaft für Ur- und Frühgeschichte*, DGUF) in Aschaffenburg in 2010 – dedicated to the theme 'Archaeology and Power' (*Archäologie der Macht*) – provided insight into the working conditions and expectations heaped on Macedonian archaeologists. Archaeological sites and finds should now back up the claim to a Macedonian heritage, he said. They had to be interlinked with the glorious Macedonian past somehow, and especially with Alexander the Great. So a weathered statue of Aphrodite would be re-interpreted as a Statue of Alexander himself and all interpretations were being subdued to a nationalistic narrative. Thus, while starting from a defensive position, the Macedonian claims to Alexander become more and more offensive. In other words – exactly what was feared by the Greek state to begin with. In this case, telling the (wrong) story and misusing archaeology for a nationalistic project of identity building becomes more and more of a self-fulfilling prophecy, with both sides becoming ever more entrenched in their own arguments. These stories have proven to be powerful as well as dangerous.

As mentioned initially, all stories are constructs by definition. The case of Vergina and Alexander the Great is an obvious example of this; of how human knowledge is constructed through current aims and information coming into contact with existing knowledge gained by experiences. The stories are built around a specific place, fed with legends and backed up with real finds and classical sources, whereafter they are bent to the will of those who want to use them for their own purposes. So when scientists use stories to bring their findings to life and to communicate them to a wider public, they have to understand that they themselves are constructing these stories. Constructions can be analysed (deconstructed) to understand their intentions and workings. They can of course also be re-assembled and thereby new stories can be created. Storytelling in archaeology is always a long process of collecting data, comparing, interpreting and finally building a model – the narrative. It is a powerful tool to better understand the archaeological finds and to give meaning to dead objects. But during the process one has to be aware of the process, of the fact that it is a model constructed and not a truth revealed.

Stories of landscapes

Narratives are also multilayered. Starting with the fundamental geological conditions of a region one can construct a narrative explaining the regional style of building, the economic and technological developments and social changes. The red sandstone of the Spessart and Odenwald mountains makes for a perfect building material. It is both malleable and quite durable, and easy to break into big building blocks. The river Main offers an easy way of transport for the bulky material, why the major quarries are lined up along its banks, especially at places where the steep cliffs provide easy access to the sand stone in direct vicinity to the river and can be transported directly. The great buildings constructed from this material are lined up like pearls on a chain along the river banks: from the great castle of Aschaffenburg, towering like an artificial mountain over the Main valley, to the imperial high church of Frankfurt and the central place, the 'Römer' – arena for the crowning ceremony of many German Emperors – to the mighty cathedral of Mayence. The representative buildings, with their deep red colour, are a dominating feature of the region and an important part of the regional identity.

The sun drenched steep hills of the river bank have also fostered the cultivation of wine yards, and the Frankonian wines of the region are still famous. But in

Bürgstadt the wine yards in the immediate vicinity of the great quarries have led to a very specific production of table grapes. These grapes, bred not for making wine but for eating, were a delicacy which could be shipped at low cost and good speed on the regular boats running from the quarries to Frankfurt. There they could be sold at a very good price to the rich burgers at the markets and fares. For the better part of two millennia, the whole system depended on the river as a transport route for bulky goods. But with the invention of the railway, and the following construction of a dense railway system in the 19th century, the importance of the river as a road declined. Quarries could be opened or enlarged along the railway lines in the inner Spessart, especially close to the stations. So a complete new sand stone industry developed in the Spessart, like the quarry in Heigenbrücken, one of the major stations along the new railway with a villa built by the successful entrepreneur, made of local sandstone of course. Soon the social structures in these villages and market towns changed, with new entrepreneurs emerging who were able to seize the moment.

On the other hand the railway also meant the eventual decline of other commerce. The Frammersbach teamsters for example, already in decline, were completely replaced by the railway. Frammersbach, which was bypassed by the railway, lost its importance while the neighbouring Partenstein profited from the change as it became a railway station. A heavy spade mine could be opened as the heavy material (not too valuable in itself) could now be transported out of the Spessart in a cheap way. All taken together, one could say that all upland regions in Europe actually lost out during the Industrial Revolution. Rich in poor sources, the Spessart was a good economic landscape in medieval and post-medieval times, with iron, silver, copper, salt, wood and many brooks and small rivers where mills could be established. The driving force of medieval industry provided an ideal basis for a pre-industrial economy, but when the industrialisation began these sources proved too poor to support the modern industries. On the other hand the railway did not only provide a cheap means of transport for cheap but bulky goods out of the Spessart, it also allowed for the import of ready-made cheap products from the industrial centres (for the story of the socio-economy and industrialisation of the Spessart, cf. HIMMELSBACH 2006; ERMISCHER 2010).

Stories of caution

So far telling the story has allowed us to draw a vivid picture of a region, its development, the interaction of basic geological and geographic factors, economic, social and technological changes. The sites and finds have become part of this picture, come to life in ways both interesting and understandable. When direct parallels can be drawn to modern developments, like the technological changes which led to the decline of a region but still offered chances for daring entrepreneurs, the picture gets really vivid. If one looks at the political development in Germany in the 19th century, one can also see how the many borders, tolls and taxes raised on the routes of transport in a highly fragmented Germany served as a bulwark for the local economic circles in the Spessart area. As an outcome of the German unification process, starting with the reform of the German states during the Napoleonic wars, followed by the formation of a toll union, the standardisation of measures and finally leading up to a united Germany with one currency and no borders at all, this protective factor of the borders disappeared. Here it is easy to make comparisons with the development of the EU, the modern toll union, the dismantling of borders and the Euro as a common currency. Given the great *Angst* connected with this process of unification, as well as the general parallels of globalisation with the 19th century, they could be used to paint a very colourful picture of high emotional value. But at that point the story becomes highly political and ideological and the temptation to use it for modern propaganda increases, whether it is for or against increased European integration or globalization.

Just how tempting historic narratives can be as instruments of propaganda can be seen in both the attempts of Council of Europe and of the European Union to paint a positive picture of the European unification process. The first campaign based on cultural heritage that was carried out by the Council of Europe in the 1990ties even used archaeology as a testimony for the common European heritage. It was called 'The Bronze Age - the first Golden Age of Europe'. Given the golden colour of bronze and the many marvellous gold objects discovered from the Bronze Age, this title was brilliant. It also referred to the Golden Age, a mythical era of perfect harmony, implying that after this "first Golden Age" a new Golden Age of Europe was now approaching. One can argue about whether the Bronze Age cultures really constitute a common European cultural heritage, but in the context of this campaign the narrative was set and clearly outlined. The campaign was analysed in this capacity by Herdis

Hølleland, the results published digitally by the University of Oslo (HØLLELAND 2008). Other campaigns followed, for example ones building on Charles the Great as the father figure of a unified Europe.

The EU has coined its slogan for Europe 'Unity in Diversity'. A slogan which shall also be backed up by a historic narrative, with the help of EU funding programmes dedicated to research unearthing the common European heritage. Those within the EU who are aware that a unification process cannot only be based on economic and financial issues see history and cultural heritage as the main source for a European identity. Therefore they were ingrained in the Maastricht treaty in 1992, transforming the European Economic Community into the European Union, and the treaty of Lisbon in 2009 which serves as a surrogate constitution for the EU after the process of creating a fully fledged constitution faltered. Many books and booklets have been published by governmental institutions on the theme of a European identity created out of history, like *Europe: Values – Paths – Perspectives*, a book distributed for free by the Press and Information Service of the German government in the millennium year 2000 (THIEDE 2000). To use history and cultural heritage as a resource for building strong narratives, so desperately thought by the European leaders and administrators, is a great chance and a great risk at the same time. Both past and recent experiences point out the dangers inherent in this process, especially when it comes to how easily archaeological and historic narratives can be misused. One does not need to go back to the extreme examples of Nazi archaeology and Nazi history to make this point, but as I have tried to show through the example of Greece and Macedonia, there are many recent examples of nationalistic narratives based on archaeology and history.

Nationalistic narratives are a curious beast – constructs to support a construct. They often serve claims of superiority by one group over another, as well as territorial claims: "because our ancestors settled here already thousands of years ago, this is our country". In this manner they easily bridge the gaps of millennia, projecting modern border lines onto prehistoric societies, thereby justifying their aggression. They happily deal with the palpable inconsistencies along the lines of: "if it does not fit the frame it does not exist" or "if it does not fit we will just have to make it fit". A curious example borrowed from a co-worker may serve to illustrate this mind-set: A colleague of mine was participating in a conference in Romania, where he was regaled by a representative of the Romanian government with stories of the glorious Dacian past of her country. At one point she explained that the modern Romanians love for their national dishes made from maize was a sign of the country's unbroken traditions since Dacian times. My colleague ironically remarked that based on that reasoning, perhaps the prehistoric Dacians were actually Indians and came from America. He of course referred to the fact that maize (or corn) is an American plant introduced to Europe in the 16th century. In fact, it did not even reach South-East Europe directly, but via a detour through the Muslim world and the Turkish conquerors – obviously not a story fit to promote Romanian national dishes to Western European guests. This story may seem amusing at first, but it shows how far and how effortlessly nationalistic narratives can stray from historic reality and how they can be appropriated to fit edicts of national pride and ideology.

One possible course of action is to counteract negative narratives with positive narratives, although, negative and positive are of course always a subjective evaluation. In our society we agree (at least for the most part) that nationalistic narratives are negative and, personally, I would dare to call narratives promoting European identity building positive – even though many might question that stance. So at last we arrive at the very place we started, with the fact that narratives and stories are constructs and have to be viewed, evaluated and handled as such. When declared absolute truths they may become harmful to certain groups in society or merge with the realm of fairy tales – to quote Shakespeare´s Macbeth "full of sound and fury, signifying nothing". And we all know how that story ended.

References

CARLIE, LENNART – ERIK ROSENGREN (2003), *Europeiska landskap*. Heide.

ERMISCHER, GERHARD (2010), Early industrialisation in a typical German upland region. In: Paul Belford – Marilyn Palmer – Roger H. White (eds), *Footprints of industry. Papers of the 300th anniversary conference at Coalbrookdale 2009*. British Archaeological Reports British Series 523. Oxford. 27–42.

ERMISCHER, GERHARD – GERRIT HIMMELSBACH (2009), Das archäologische Spessart-Projekt. Ein Beitrag zur Kulturlandschaftsforschung und Identitätsbildung im Waldsassengau. *Berichte des Zentrums für Regionalforschung der Universität Würzburg* 2009.1, 165–180.

ERMISCHER, GERHARD – HARALD ROSMANITZ (2007), Building cultural routes to develop a marginal landscape: a case study from the Spessart. In: Council of Europe, *Fifth meeting of the workshop for the implementation of the*

European Landscape Convention: Landscape quality objectives: from theory to practice. European Spatial Planning and Landscape 84. Strasbourg. 309–319.

HARTMANN, GUIDO (1935), *Der Spessart in der Literatur*. Aschaffenburg (2nd ed.).

HAUFF, WILHELM (1827), Das Wirtshaus im Spessart. In: *Märchen-Almanach für Söhne und Töchter gebildeter Stände auf das Jahr 1828*. Stuttgart [here used: Hans Christian Andersen – Lorenz Frölich – Vilhelm Pedersen (2005), *Sämtliche Märchen*. Düsseldorf. 190–328].

HIMMELSBACH, GERRIT (2006), Wirtschaftsgeschichte in einer Einöde. Die Entdeckung der Kulturlandschaft Spessart. In: Hans Peter Baum – Rainer Leng – Joachim Schneider (eds), *Wirtschaft, Gesellschaft, Mentalitäten im Mittelalter: Festschrift zum 75.Geburtstag von Rolf Sprandel*. Stuttgart. 109–131.

HØLLELAND, HERDIS (2008), *The Bronze Age – the dawn of European civilisation? A case study of usages of the past within the European identity discourse*. URN:NBN:no-20976. (https://www.duo.uio.no/handle/123456789/23200 [26.07.2013]).

KAMPFMANN, GERHARD – STEFAN KRIMM (1988), *Verkehrsgeographie und Standorttypologie der Glashütten im Spessart*. Studien zur Geschichte des Spessartglases 2. Veröffentlichungen des Geschichts- und Kunstvereins Aschaffenburg 18.2. Aschaffenburg.

KRIMM, STEFAN (1982), *Die mittelalterlichen Glashütten im Spessart*. Studien zur Geschichte des Spessartglases 1. Veröffentlichungen des Geschichts- und Kunstvereins Aschaffenburg 18.1. Aschaffenburg.

KRIMM, STEFAN (1984), Zur Geschichte der Waldglasproduktion im Spessart. In: Claus Grimm (ed.), *Glück und Glas – Zur Kulturgeschichte des Spessartglases*. Veröffentlichungen zur Bayerischen Geschichte und Kultur 2. München. 159–179.

LOIBL, WERNER (2001), Der hessische Glashandel in die Niederlande im 17. Jahrhundert. *Nassauische Annalen* 112, 75–125.

MACAULEY, DAVID (1979), *Motel of the mysteries*. New York.

MOSER, PETER (1990), *Mittel- und Nordwesteuropäischer Landtransport: Die Frammersbacher Fuhrleute und ihr Beitrag zur Transportgeschichte (15.–19. Jahrhundert)*. Bamberg.

PFISTER, LUDWIG (1812), *Aktenmäßige Geschichte der Räuberbanden an den Ufern des Mains, Spessart und im Odenwalde*. Heidelberg.

PRATCHETT, TERRY (1997), *Hogfather*. London.

PRATCHETT, TERRY (2005), *The science of discworld III: Darwin's watch*. London.

QUACK-MANOUSSAKIS, REGINE (2003), *Die deutschen Freiwilligen im griechischen Freiheitskampf von 1821*. Jahresgabe des Förderkreises König-Otto-von-Griechenland-Museum der Gemeinde Ottobrunn 5. Ottobrunn.

ROSMANITZ, HARALD (2008), Auf den Spuren des Spessartglases – Die archäologischen Untersuchungen auf der Burg Bartenstein bei Partenstein. In: Helmut Flachenecker – Gerrit Himmelsbach – Peter Steppuhn (eds), *Glashüttenlandschaft Europa: Beiträge zum 3. Internationalen Glassymposium in Heigenbrücken/Spessart*. Regensburg. 84–92.

ROSMANITZ, HARALD (2011), The castle project in the Spessart – scientists and volunteers explore a cultural landscape. *Concilium Medii Aevi* 14, 101–120.

THIEDE, CARSTEN PETER (ed.) (2000), *Europa: Werte – Wege – Perspektiven*. Ratschläge und Hinweise. Berlin.

VON BERLICHINGEN, GÖTZ (1562), *Lebens-Beschreibung des Ritters Götz von Berlichingen zugenannt mit der Eisernen Hand*. Ed. by Engelbert Hegaur (1910). München.

VON GRIMMELSHAUSEN, HANS JAKOB CHRISTOFFEL (1669), *Der abentheurliche Simplicissimus Teutsch*. Nürnberg. [here used: Bibliothek der Weltliteratur. Essen – Stuttgart 1985].

WEDEPOHL, KARL HANS (2003), *Glas in Antike und Mittelalter. Geschichte eines Werkstoffs*. Stuttgart.

Archaeology as European Added Value

ELISABETH NIKLASSON

Abstract:
Ever since the Council of Europe's campaign 'The Bronze Age – The first Golden Age of Europe' was carried out in the 1990s, archaeologists have from time to time expressed concern about periods or interpretations being used as tools for forging a collective identity tied to the geographical, political and ideational entity of 'Europe'. While narratives of a pan-European unity in prehistory have often been disregarded or deconstructed by the research community, discussions on increasing cross-border cooperation and harmonization of practises is still on-going. Both of these directions warrant continuous critical consideration, especially since it is suggested that a future 'Archaeology of Europe' rely in part on the opportunities for financial support offered by the European Commission. European Commission enabling structures, especially within the field of culture, still talks of a common European past. By focussing on 'European added value', the first award criteria for achieving funding under the European Commission Culture programmes, this paper discusses what meanings such a value evoke in relation to archaeology and cultural heritage. It also provides some examples of how such meanings can become visible in the narratives of co-funded archaeological projects.

Archaeological narratives (grand or petit), carefully moulded out of archaeologically produced facts, are not created in a vacuum. They are written in a specific time, under a specific set of circumstances with a specific set of objectives in mind, all of which assist in their formation. For the most part we recognize (at least principally) that embedded 'plots' – employed to provide our narratives with a sense of "continuity, coherency, and meaning" (WHITE 1987: 11) – do not arise spontaneously from the isolated mind of the enlightened researcher, but are contingent on discursive environments and inherent modes of storytelling. We also know that ordinarily, before we can even start to build an empirical base permitting us to tell a story which claims can be evaluated by peers, we need some sort of basic means or economic incentives. These prerequisites are intrinsically bound; archaeology cannot survive if no one is interested in funding it, and if our narratives do not correspond in some way to lived experiences or political strategies in the present, they risk generating little or no economic support. This is all very well, but I hardly need to call attention to the importance of funding among fellow archaeologists. Money, or rather the lack thereof, is a topic discussed on a daily basis in most branches of the discipline. However, when it comes to the funding arrangements themselves, as crucial as they are to any archaeological undertaking (or to actions in any other field for that matter), they are rarely seen as anything more than an annoying circumstance or unfortunate necessity. And, at best, they are something signified by a logo or an acknowledgement at the end of a publication. Contrary to such views, I believe that the conditions underlining such arrangements is of great importance, and a very good place to start if we are interested in increasing our knowledge about the 'values' or 'political economy' of archaeology, and to raise the state of critique.

For that reason, this paper is centred around the first criteria applicants must fulfil in the European Commission funding programmes Culture 2000 and Culture 2007–2013, the one of 'European added value'. The criteria under which financial support is granted is one of the main points at which benefactor and beneficiary meet, interact, oppose and become part each other's storylines. By using examples from archaeological projects which have benefited from co-funding from these programmes, I will look at whether and (if so) how this notion of 'European added value' can be perceived in relation to their plots and narratives. But first the criteria itself and the choice of circumstance, which raises questions about the aforementioned issues of discursive environments and inherent meta-structures, calls for a discussion on what the 'European added value' of archaeology might be to the European Commission. Turned around, it also raises questions about the value of Europe might be to archaeology. In other words, to what extent do these funding bodies by way of (geo)politically defined criteria like European added value, and alongside well-known regulating mechanisms like peer-communities, interact in the creation of archaeological narratives and infrastructures? And, since archaeologists (while increasingly reflexive and by no means naïve) do not enter into discourses on Europe as blank slates, how might pre-set ideas of Europe be induced or wilfully offered when narratives are written? Drawing upon documents, texts and ethnographic material collected during my on-going PhD studies, I will engage with these questions from three angles; perceptions of funding in archaeology, archaeology and EU/European added value, and lastly, European added value and archaeological projects.

On funding

All archaeology involves funding of some kind and all types of funding involve politics on different levels, both in a practical sense and as part of the infrastructures of power at work in all human interaction. It needs to do so if we accept that firstly, archaeologists should get paid for the wide range of tasks they perform. Secondly, that we are not practising archaeology only for ourselves but to contribute to society in some way. And thirdly, that regardless of our approval, archaeological narratives are still drawn upon in politics and when building collective identities, so we had better participate if we want to make our voiced heard.

Despite being so central, the conditions under which funding is granted are seldom explored in a critical fashion. Do not get me wrong, the various financial ties between archaeologists and state authorities are often assumed or referred to implicitly when dealing with the range of themes related to archaeology and modernity (e.g. DIAZ-ANDREU 2007; KOHL – FAWCETT 1996; THOMAS 2004; TRIGGER 2006 [1989]), but the practicalities of these funding arrangements are rarely developed upon in any detail (for exceptions see HAMILAKIS 1999).

Meanwhile, studies looking specifically at financial aspects of archaeology, aiming to situate the profession in a larger socioeconomic context (see e.g. AITCHISON 2000; 2009; AITCHISON – SCHLANGER 2010; DEMOULE 2004; 2011; WILLEMS 2009), have mainly been concerned with describing national financial structures and assessing the consequences of certain types of funding (such as commercial- developer- or state funding), rather than reviewing aspects of political agendas or their ideological roots. Importantly, the gap between archaeology as Cultural Resource Management and academic archaeology has been addressed in this context, although the main concern seems to lie with the level of scientific thoroughness and the overall degree of knowledge production, and less about the potential cognitive impacts or how funding agencies might play a part the content and character of the questions posed.

One reason for the lack of interest in such aspects could be the prevailing view of archaeological research and practise as detached – at least intellectually – from the political aspects of the everyday work (i.e. legal permits, administrative procedures or funding engagements). As emphasised by Michael Shanks when talking about the political economy of the discipline of archaeology, politics is still seen as the context in which archaeology takes place, or how the results are used, rather than being an integral part of the practice (SHANKS 2004:

497). In this view archaeology is often seen as a science based on a "stock of knowledge embodied in theories", thereby expelling funding issues from the arena of rationality and criticism and reducing them to an outside interference, getting in the way of researchers autonomy (ADDELSON 2003 [1983]: 179). This makes inquiries into any proposed interaction between the two seem redundant at best. After all, what you write in an application is just prose, having no real bearing on what you actually do right? In line with Shanks I would argue that this relationship goes beyond cosmetics and that we need consider all parts of archaeological practice as bound to its epistemology. Even the most basic declaration, that funding creates better opportunities for researchers within the dominant traditions to exercise 'cognitive authority', and help others of their own persuasion to rise in the ranks – thereby gaining even more funding – shows how funding effects the contents and organization of knowledge production at any given historical moment (ADDELSON 2003 [1983]: 177f). Ultimately, it affects how we will come to understand the past and present (see ADDELSON 2003 [1983] for in depth discussion). Viewed in this way, funding engagements becomes something we cannot afford to dismiss, and a way to explicate current entanglements with discourses of power.[1]

Of course, archaeologists are no distinct species and this discussion could apply to almost any field trying to increase their status as well as our knowledge of the world. Still, each field of thought presents its own characteristics and challenges. In a study on peer review panels in the United States, social scientist Michéle Lamont found that the notion of 'excellence' was defined very differently depending on the field of the evaluator, and that "disciplines shine under different lights [...] and are best located on different matrixes of evaluation, precisely because their object and concerns differ so dramatically" (LAMONT 2009: 9). Furthermore, the historical underpinnings of archaeology as a state authorized activity, legitimising heritage in the construction of national pasts, makes it worthwhile to consider in its own right. With this in mind, EU funding programmes in culture stand out as particularly thought-provoking when it comes to the fields of heritage and archaeology. This is partly because it represents a relatively new level for cooperation, with potential effects on disciplinary infrastructures, and partly because it can act as a source of inspiration for potential reincarnations of longstanding notions of "European-ness".

---

1 In this text I refer to discourse as being simultaneously a piece of text, an instance of discursive practice, and an instance of social practice; the way social and political domination are reproduced in text and talk (FAIRCLOUGH 1992: 4; 1995).

European added value – The concept

> "The conversation (about European Added Value) is constant. In everything the commission does it is always about: is this something Europe should deal with? Is it something we do better here?" (EC-EAC03 2012, European Commission official, Culture, translation mine).

Before going any further, it might be helpful to know a bit more about this concept. Ever since my time working as a trainee for the European Commission, its vagueness has intrigued me. As part of the team of administrators managing the selection process for the programme Culture 2007–2013, one of my tasks was to assist experts and when needed, clarify how certain criteria should be addressed. This was not an easy task and with experts from a range of different countries and fields, with varied experience in evaluation procedures, the differences and nuances in how criteria were approached fascinated me. European added value (from now on EAV) is the first listed amongst the three highest valued award criteria, declaring that projects supported by Culture 2000 needed to "provide real European Cultural added value" (Call for proposals 2003), while projects supported by Culture 2007–2013 are judged based on "the extent to which the project can generate real European added value" (Programme Guide 2010).

Used as a condition for most EU programmes and actions, and embedded in a family of interrelated concepts,[2] the notion has become hugely popular during the last decades and anything having to do with the EU agenda now come equipped with the claim that the proposed actions should lead to an increase of such a value. As pointed out by political scientist Daniel TARSCHYS (2005: 37), the extensive use of the argument almost vanquishes its potential credibility. Of course, the meaning of the notion of 'value' itself is not easy to delimit, and adding European to it makes it even more complex. From a social theory point of view, anthropologist David Graeber argues that there are mainly three ways in which 'value' has been used; values in the sociological sense of what is good, proper, or desirable in human life, value in the economic sense as measured by how much others are willing to give up to get desired objects, and value in the linguistic

---

2 As a consequence of what has been called 'The value Turn in Governance and public policy', many similar concepts like Surplus value, The value chain, Common values, Nordic value or "*Nordisk nytta*" have developed (TARSCHYS 2005).

sense as 'meaningful difference' (GRAEBER 2001: 1f.). All three understandings have bearing on the concept of EAV.

According to the EU Council resolution establishing Culture 2007–2013, "European added value is an essential and determining concept in the context of European cultural cooperation, and a general condition for Community measures in the field [...]" (2006: 1855/EC). Such declarations of importance are not necessarily followed by explanations of *how* it is important or *what* it stands for, although an attempt to nail down its content was made in 2003. Then it was stated that it should apply to: pan-European, multilateral, cooperative, visible, knowledge generating and awareness raising actions – primarily benefiting European citizens – so as to create a sustainable economic development and integration of cultures (2003: C13/03). In other words, a sound economic justification should be matched with positive contributions to a sense of European unity, making EAV about what is "good" and "right", as well as the worth of cultural actions in an economic sense. It is also stated in the document, similarly to the UNESCO notion of 'Universal Value' (see CLEERE 2001), that EAV is meant to "be implemented in a flexible way", leaving it defined by its vagueness, simultaneously empowered and undermined by it. In a linguistic sense it is perhaps the word European that is most significant. Notwithstanding that both the geographical delineation and adjective 'European' are epistemological entities (with ontological effects), the notion is laden with meanings (progress, Christianity, enlightenment, democracy, colonialism, totalitarianism etc.). Since the majority of these embedded meanings – persistently invoked in EU rhetoric's –[3] builds on notions of "us" and "them", of inclusion and exclusion, it is hard to imagine how a flexible European added value (note that value is in singular, not plural) could do anything but reinforce division. While all types of identity build on division, which is not by default a negative feature, there is a risk that already existent western chauvinism settles comfortably in this mould. Among the effect these rhetoric's have already had, the now habitual conflation of 'Europe' with the space of the European Union can be mentioned. Tanja Petrovic has called this "thinking Europe without thinking", explaining that countries "[...] already within the EU can include or

---

3 As a recent example, European Commission President J. M. Barroso, said in a speech at the opening of the archaeological and historical exhibition 'Europe meets the world' at the National Museum of Denmark in January 2012, that "We are always speaking today, because of the crisis, about the financial markets, but Europe is about culture, Europe is about values, Europe is also about civilization [...] we should not be afraid of the word 'civilization'" (BARROSO 2012, original language).

exclude, while those who are 'on their way to Europe', or those who do not have the option of membership at all, are excluded" (PETROVIC 2011: 1).

Discussing the works of Clyde Kluckhohn, Graeber explains his definition of value as "conceptions of the desirable", in the sense that the "desirable refers not simply to what people actually want [...] Values are ideas about what they *ought* to want" (GRAEBER 2001: 3, original emphasis). This could be a good way to approach the concept of EAV, as being about what those working towards an "ever closer union among the peoples of Europe" (1957 Treaty of Rome) want EU citizens to want. To be fair, and temporarily disregarding the 'democratic deficit' which has plagued these institutions, it also comes down to what the voting publics want. Either way, it is not just a wish or mere abstraction. The concept of EAV affects the way people in the EU do things, as well as the evaluators and participants in funded projects who have chosen to relate to it.

As a final remark, EAV is certainly not the only vague criteria that experts and applicants have to interpret (others like 'intercultural dialogue' and 'synergy effects' can be even more difficult), but it is one of the core concepts in all actions of the European Commission and as such it becomes interesting in relation to archaeology and cultural heritage, both due to the symbolic and economic role heritage have been given in European cultural politics, and due to archaeology's role in authorizing national pasts.

Archaeology as European/EU added value

Now let us consider this connection between the European Commission and activities in the field of archaeology and heritage, by looking into the premises for their respective investment in each other. Namely, wherein lays the European added value of archaeology and heritage for the European Commission (if anywhere)? And what might European added value be in relation to archaeology?

To begin with, in the context of European Commission actions, archaeological activities and sites are usually included into the wider category of cultural heritage and have only occasionally been discussed explicitly or on their own.[4] Consequently, it is as one of the essential components of the set phrase 'European

[4] However, as discussed by SMITH (2006) the European or western 'authorized heritage discourse' – a discourse favoring monumentality, scientific/aesthetic judgment, nation building and institutionally sanctioned knowledge – does tend to privilege archaeological conceptualizations of heritage. This is reflected in most EC and EU conceptualisations of heritage.

Heritage' they have become activated and meaningful (see TZANIDAKI 2000 for an extended overview). In this shape, there have been mainly three types of added value assigned to them; intrinsic European value, value for contemporary European culture and identity and commercial value (i.e. tourism, employment, PR and branding).

Already in the Treaty of Rome (1957), establishing the European Economic Community (EEC), the protection of "national treasures of artistic, historical or archaeological value" was mentioned (Art. 36). But it was not until the 1970s and the 'Declaration on European Identity' (1973) – in which "reviewing the common heritage" was listed as the first measure – that heritage was put to work in relation to goals of cohesion. Up until then the EEC had relied on federal-functionalist theories of integration, in which the integration of individual economic sectors was expected to generate spill-over effects. These effects would then translate into cultural cohesion and an ever stronger role for to supranational institutions (see ROSAMOND 2000). In the 1970s the EEC realised that this idea of a European ethos or common identity, emerging by itself as a bi-product of financial commitment, was not likely occur without directed efforts. The new message, as it was passionately expressed by a French representative of the Council of the European Communities at the time, became: "There is no political power without economic power. There is no economic power without political and cultural purpose" (1981: 14/11/CEC, also in TZANIDAKI 2000). The preservation of heritage was included in this purpose, and the motivation for supporting it was articulated in the first Recommendation on the topic: "…the architectural and natural heritage which reflects Europe's cultural identity is seriously threatened with decay and disappearance and urgent measures are needed" (1974: 75/65/EEC). This emphasis on heritage in a state of crisis, was repeated in recommendations and resolutions throughout the 1970s, and focus was placed on saving and preserving the common tangible heritage, as well as contributing to an increased workforce of heritage experts and improved techniques, thereby increasing tourism and movement of cultural goods (1977: 6/77 EEC). The fear of loss and the destruction of sites and monuments was a real concern in the post-war period (and still is), and it was believed that the social and cultural values they represented would also wither in their wake (SMITH 2006: 100). As a result, their intrinsic European value and their potential value for a European identity, was dependent on the discourse of a heritage in crisis, passively waiting for someone to rescue it for the benefit of all Europeans.

Alongside its economic value this inherent value was emphasised throughout the 1980s, but with a twist towards a more positive notion, accentuating its dynamic cohesive properties. In relation to the European Parliament 'Report on the protection of the architectural and archaeological heritage' (1982), a report which led to financial support for projects on sites like Skylletion (IT), Acropolis and Mount Athos (GR), the motivation for Community involvement was expressed in the following way:

> "No aspect of culture gives stronger and clearer expression to the link with the past than architectural monuments and archaeological sites. They are the silent but revealing testimony to our European cultural history, because monuments have a European character. They are not national, styles from ancient times to the present being European. They therefore kindle the awareness among Europeans that, despite the great variety, they have a common culture, and this, even more than a common economy, is what is really needed for a united Europe" (1982: 0206 EP).

Rhetoric's aside, it is important to note that what little financial support came from the EEC at this time was generally connected to cooperation with the international (non-EU) organisation Council of Europe and with UNESCO. It was not until 1992, when the Maastricht Treaty broke new grounds by establishing a legal base for cultural action (1992: Treaty of Maastricht) – and the European Communities became the European Union – that the European Commission could carve out their own space in the field of culture. While there were (and still are) strong subsidiary rules in place, stating that the European Commission could not in any way interfere with national interests in culture, the first funding programmes in this field were launched only a few years later. Among them were the Raphaël programme (1997–1999), which was entirely dedicated to heritage and engaged archaeologists and heritage professionals on many different levels (see NIKLASSON 2013). The first generation of programmes were followed by the interdisciplinary programme Culture 2000, and the current successor Culture 2007–2013, both supporting heritage activities as only one out several areas. While these programmes only offer co-funding for projects, and even though the scale of investments cannot be compared to ones in other fields and sciences, it is important to recognise that cultural actions have long served as food for commissioners and politicians longing to anchor the myth of European Unity in a past a little bit more distant, a little less dark (see RIEKMANN 1997).

So, what happened to the European added value of heritage when these programmes came into being? With the famous Maastricht Article 128 – stating that

> "the Community shall contribute to the flowering of the cultures of the Member States, while respecting their national and regional diversity and at the same time bringing the common cultural heritage to the fore"

– cultural heritage slowly came to be seen as something more malleable and mobile, as an active agent in identity building. The presumption that it was something already in existence that could be brought to the fore, was now accompanied by the view that heritage activities with a 'European dimension', could increase the European added value of monuments and sights (1997: 2228/EC). As an expression of this non-static value, it was stated in the webpage of the Directorate for Culture and Education in 2004 that: "cultural heritage is widely recognised across Europe as a vehicle of cultural identity".[5]

Today, although the notion of heritage is still heavily linked to the fields of archaeology, preservation and the museum sectors – intangible heritage, diversity and socioeconomic values, are more and more emphasised in cultural actions.[6] In an interview with a European Commission official in Culture, upon the question of what kind of attitudes there were towards heritage in the commission, he concedes that:

> "It's a hard topic, or a very traditional topic […] the fact that the [Culture] programme has a multidisciplinary approach is good in that respect cause it is not focusing on […] too traditional aspects

---

5 http://ec.europa.eu/culture/portal/activities/heritage/cultural_heritage_vehic_en.htm [28.06.2013].

6 This change was highlighted in the rhetoric of EU enlargement, a long process which led to the re-definition of heritage towards intangible values like "democracy, pluralism and the rule of law" (TZANIDAKI 2000: 23). Another important influence has been the UNESCO Convention on the protection and promotion of the diversity of cultural expressions, adopted on the 20th of October 2005 (2005: UNESCO). It aims to preserve cultural diversity worldwide and the Community became party to it in 2006.

of culture but it can be transversal so there are projects that are born as more traditional ones for example with activities focusing just on heritage and then they found their way and linked to other sectors" (EC-EAC01 2011, original language).

In this context 'traditional' was later exemplified as restoration or public heritage activities, the latter being something archaeologists would probably not consider in this way.

The need for intersectional cooperation is often stressed in relation to the economic value of heritage. Mirroring the general political and economic situation in the EU and elsewhere, the emphasis on economic value has increased drastically in recent years (especially in relations to the coming programme *Creative Europe*). When asking the same official what this might mean for the value of heritage I got the response:

> "I participated in a few seminars and debates organised by actors in the field of heritage and [...] they asked also for some contributions on the classic value of heritage if you can call it that, or social value. I mean there are different factors and all of them are equally important of course, but probably nowadays the economic one is the most visible [...] [but] let's not forget that heritage is quite a complex concept so it is not just economic and that is something we always try to stress" (EC-EAC01 2011, original language).

This ambivalence is widespread in EU discussions involving cultural heritage. While the economic value has always been acknowledged (especially in relation to tourism and the heritage industry), the ideas of an inherent European value residing inside monuments versus a dynamic value as building blocks for a European sense of belonging, have never successfully merged. This parallel existence, in the field of tension between *promoting* and *creating* a heritage for Europe, where well-seasoned meta-narratives can be used together with a freshly assigned European significance, can be compared to the writing of a book after the movie is already made. As anthropologist Cris SHORE (2000) has observed, European Commission discourses on Europe have their roots in nineteenth century liberal modernity, and the institutions have, at least until the beginning of the 21st century, used the same methods for identity building as the nation states (flag, hymn, currency, heritage, specific days). According to him this builds on the flawed assumption that a European identity can be created by tapping into already existing patterns of 'European culture' and 'core values'. That such an identity will emerge through stimulating the awareness of a shared cultural heritage, all while ignoring the fact that precisely those cultural elements have stood as dividing walls between Europeans (SHORE 2000: 225). Shore, along with other anthropologists and social scientists, agree that the initiatives undertaken so far have not been very successful (DELANTY 2010; SHORE 2000; 2006; THEILER 2006). The European Commission has managed to "Europeanise" some of the cultural infrastructures within its jurisdiction, by offering a spatial category outside the realm of national authorities and reconfiguring national elements into "European" products (TRETTER 2011: 945). Yet, top-down initiatives looking to generate a European added value, by prescribing what EU citizens *ought* to desire, does not seem to have had the anticipated impact on a societal level.

Roel DURING (2010) and Claske VOS (2011), two researchers who have separately looked into how EU discourses on heritage have been transferred and negotiated on local and regional levels, both insist that the gap between those active at a policy level and those implementing activities at ground level, has been too wide. This has led to a great deal getting intentionally or unintentionally lost in translation, and their results show clearly that what becomes 'European' in the set phrase European heritage (if anything), is not always what was originally hoped for. For example, the Regional Programme on Cultural and Natural Heritage in South East Europe, funded by the EU and the Council of Europe in an attempt to bring about Europeanisation in applicant states, ended up working contradictory to their goals of reconciliation in Serbia (VOS 2011). According to Vos, the programme created a division between the people who were for and against the initiative, and the sites selected for support became the ones the national institutions lay no claim to or had no interest in. That they did not succeed as economy boosting, goodwill-generating instruments does not mean they had no effect. While publics may never have understood that the annual heritage day in their country is an EU initiative,[7] or cared much about the new European dimension of heritage sites in Serbia, archaeologists and other professionals in the field of heritage are more connected on an EU scale than ever and this is bound to have an effect on the development within our fields.

[7] Similar to the examples discussed by Claske Vos (2011), European Commission officials have made some remarks in my interviews, that the event called the European Heritage Days tend to become 'nationalised' in countries where it is a successful initiative, while the event is still called 'European' in places where it is not.

As a final remark it might be wise to say a word or two about the future of heritage in the European Commission. Despite or perhaps because of the current European sovereign-debt crisis (starting in 2008), the role and significance of cultural heritage is being renegotiated. So is the support for humanities and cultural actions within framework programmes dealing with Research and Regional development (from which cultural heritage projects have benefited greatly). When the Green Paper on 'Horizon 2020', the future major EU funding instrument for research, was presented in February 2011 it immediately became clear that research into subject areas like archaeology and heritage – along with other topics in the social sciences and humanities – was not to be considered a priority or hardly even a concern. This led to strong objections from various disciplines and institutions. While debates continue, the European Commission's directorate for education and culture has adopted – alongside already supported initiatives such as the 'European Heritage Awards' and the 'European Heritage Days' – its perhaps most emblematic heritage initiative as of yet, the 'European Heritage Label'. This new label will focus especially on young audiences and on Europeanising the didactic contents on selected sites, rather than restoration, popularity, aesthetics or cultural historical values (2011: 1194/EU). On the whole it appears as if heritage is still (or perhaps to an even greater extent than before) seen as something that needs to be highlighted, preserved and communicated within EU cultural actions, rather than critically analysed and researched. This type of asymmetry has the potential to increase uncritical practises in regards to narratives woven around archaeological sites, actualizing the question of where in the span of tension between a future 'Archaeology of Europe' and a 'Europe' in archaeological interpretations, the notion of European added value has its place.

Europe/EU as added value for archaeology

Having talked at some length about the values of heritage within the European Commission, I would now like to narrow the scope and turn to the discipline of archaeology to discuss how this added value could manifest itself and what it could entail.

Aside from the most basic value which the programmes of the European Commission holds to archaeology, the one of economic opportunity, there are three interrelated ways in which this could be approached:

- One is the value of Europe as a frame for interpretation of the past, which in some cases have led to the search for a unique 'European-ness' corresponding to the landmass called Europe and the peoples residing there. This could be named the archaeology of a European past.
- The second is what could be called the "non"-value of Europe, as presented in scholarly efforts to deconstruct its meanings, in relation to the meta-narratives of modernity (including the accounts of a European past created in the former category). In a way, the sheer wealth of texts written on this topic during the last 30 years has in itself created a trope of Europe, an array of specific ways for deconstructing it and eventually recreating it in the process of tearing it apart.
- A third way to approach a European value is by looking at the influence of geopolitical institutions and infrastructures as enablers, thereby highlighting the potentials and the consequences of the EU as a level of governance affecting archaeological practice. It actualises the Europe of the EU as a "new" platform for cooperation. Let us call this a future archaeology of Europe.

Within Archaeological storytelling, long dominated by meta-narratives such as evolutionary theory or Marxism, Europe have figured as everything from an interpretative geographical frame – the backdrop or stage upon which the plot has played out – to starring as the main character or phenomenon to be explained, sometimes turning out to be the very explanation in itself. Late 19th and early 20th century narratives of Europe habitually followed a teleological path of progress reaching its crescendo in modernity, and whether it was due to uncorrupted barbarian origins or the resurrection of a civilized classical past, peoples in Europe were attributed with certain favoured characteristics, distinguishing them from various Others (i.e. CHILDE 1925; 1962; CLARK 1952; HAWKES 1940; RENFREW 1973). Archaeologist Herdis HØLLELAND (2008) has considered such narratives in some depth, paying particular attention to the Bronze Age and the works of Gordon V. CHILDE (1892–1957). She argues that, while archaeological grand narratives of Europe have in different ways been uncritically incorporated in the political identity discourses of the Council of Europe and the EU (especially when it comes to European personhood and citizenship), they have also had the reverse effect and contributed to an increased openness and critical awareness in Bronze Age archaeology (HØLLELAND 2008; 2010; 2011). The particular use of the Bronze Age has also been considered by archaeologist Anna GRÖHN

(2004). Although she creates a good overview of projects benefiting from support from the European Commission Culture programme (up until 2004), one main point of focus in her study – as in most texts on this topic – is the Council of Europe's campaign 'The Bronze Age – The First Golden Age of Europe' (1994–1997) (COUNCIL OF EUROPE 1994). These campaigns, followed by exhibitions like 'Europe in the Time of Ulysses – the European Bronze Age' (1998–1999), demonstrated how archaeologists interacting with ideas of Europe in a political setting, had little trouble tying together a past, present and future Europe. A campaign pamphlet from 1993 suggests that

> "The Bronze Age was an exciting period of Europe's history when contacts between the various parts of greater Europe began to develop, mirroring the modern interaction between the member states of the Council of Europe" (CoE 1993).

Similarly, in the theme number of 'European Heritage' (1994 No. 2), quotes by Gordon V. CHILDE (1930: 2) such as '...all the vital elements of modern material culture are immediately rooted in the Bronze Age', were used to tie the period to the Europe of today. These campaigns and exhibitions later received attention from both critical archaeologists and enthusiastic European Commission officials, the latter intensifying their cooperation with Council of Europe in the area of culture and cultural heritage from the mid-1990s onwards (1996: 7575 CoE). Alexander Gramsch has argued in relation to the campaigns and what he saw as a growing Europeanism in archaeology – or the hunt to "present histories of the 'Europeanness' of prehistoric cultures in Europe" (GRAMSCH 2000: 7) – that we need a 'Reflexive Theory' rooted in sociological theory and epistemology, to guide future research on prehistories in Europe, taking into consideration the fundamental differences in the archaeologies of the continent (GRAMSCH 2000: 16). While there are signs that such a change is slowly taking place, there are still traces left of what Martin Bernal has called the "[...] continental chauvinism inherent in all our historiography, or philosophy of writing history" (BERNAL 1991: 32). Even though the approach suggested by Colin Renfrew in the 1970s, of "[...] seeing the events of European prehistory as the result of a purely local process, in essentially European terms" (RENFREW 1973: 121), probably finds little or no support today, attempts to question Europe as a frame for interpretation often end up arguing for the validity of such a statement in the first place. And on that note, it might be time to enter the discussion about the so called "non"-value of Europe.

Grand narratives have received significant attention in the humanities ever since the late twentieth century's 'linguistic-' and 'cultural turn'.[8] Having corresponded particularly well to the political movements of colonialism, imperialism and nationalism, archaeological narratives is no exception (PLUCIENNIK 1999: 656). Emerging in pace with the political project of building Europe, this shift towards questioning the assumptions beneath constructed pasts (and presents) has led many scholars to ask 'when was Europe?', 'what is Europe?' and 'how is Europe done?' (BUGGE 2003; BALZARETTI 1992; BURKE 1980; 2006; GODDARD *et al.* 1994; DELANTY 1995; 2010; MALMBORG – STRÅTH 2002; SHORE 2000; PAGDEN 2002; THEILER 2006 etc.).

In the meantime, in the wake of dissolution and epistemological upheaval – situated knowledge's and imaginary communities – archaeology developed an obsession with identities. While they were now recognized as fluid both in the past and present, and no longer restricted to national borders, this also opened up for dialogues of European identities. As pointed out by Gramsch, the political reasons behind these 'turns' as they came to be adapted in archaeology, such as the crisis connected to the "dissolution of established nation-states", is at the same time partly to blame for what he calls the 'Europeanist turn' (GRAMSCH 2000: 7). Thus, although critical in nature, EU discourses on Europe have also been assisted by academic questioning and the search for 'European-ness' that intensified after the Second World War (GODDARD *et al.* 1994). By now the study of European Commission cultural policy and the notion of 'European identity' have almost become a field of its own in social and political sciences. Despite the fact that archaeologists, as pointed out by Mark Pluciennik, "are generally better at examining the past, including the history of their own discipline, than at considering the present or the near future" (PLUCIENNIK 1998: 816), quite a number of studies have considered the topic from an archaeological point of view (e.g. GRAMSCH 2000; GRAVES-BROWN – JONES 1996; GRÖHN 2004; DURING 2010; ENGLISH 2008; KRISTIANSEN 2008; PLUCIENNIK 1998; TZANIDAKI 2000; VOS 2011).

Such critical approaches (wherein I include my own research) have a tendency to become, and perhaps needs to become, traditional in their critique of tradition

[8] Deconstructed by theorists like Jean-François LYOTARD (1984) and Hayden WHITE (1987), undermined and contested by post-colonial theorists such as Edward W. SAÏD (1978) and Gayatri SPIVAK (1988).

(as in canonical). Discussing the hermeneutics of tradition, philosopher Marcia Sá Cavalcante SCHUBACK (2013) describes tradition as something inescapable that we can only approach or understand from within, as a *withsense* rather than an absence or a presence. Tradition has a critical potential in that it provides steady footing for reproach, but we cannot escape it and we keep recreating it in the act of deconstruction. In search for operational values the European Commission have both drawn on and fought traditional notions of Europe, and in midst of the economic crisis this search have also been interrupted by tradition in its various forms (like ultra-nationalism). In the critical narratives of archaeologists and other scholars, the frame of Europe may have been proven hollow and it may even have lost its image (if it was ever there). But this does not render the attempts to uphold it – the efforts to fill the frame with a new image – harmless, but all the more dubious, calling for continuous critical consideration.

This brings me to the third type of potential European added value, to the ideas of a future archaeology of Europe. While anachronistic evocations of a pan-European unity in prehistory have generally been dealt with by proponents in the former category, discussions on increasing cross-border cooperation and harmonization of archaeological practises within the EU and Europe, carry on. With the formation of the Council of Europe Conventions on archaeological heritage and the start of the European Association of Archaeologists (EAA) in 1994, and later also the Europae Archaeologiae Consilium (EAC) in 1999,[9] platforms for exchanging and integrating research and ideas, guidelines and languages became established. In 1999, seeing that concerted action was being developed and problems addressed, archaeologist Willem Willems expressed confidence in the future of European archaeology, reassuring us that "European archaeologists are working on that future, and that they are doing it together" (WILLEMS 1999: 16). A few years later, in view of new EU legislation which affected archaeological heritage, and the lack of EAA and EAC representation in Brussels, the tone was less optimistic (WILLEMS 2007: 68). Nevertheless, judging from the continued debates within the field and new platforms for cooperation – like the 'Joint programming initiative for Cultural heritage',[10] the new 'Reflection group on EU and Cultural Heritage' composed of representatives of National

[9] A network of heads of national services responsible under law for the management of the archaeological heritage in the Council of Europe member states (http://www.european-archaeological-council.org [20.02.2013]).

[10] http://www.jpi-culturalheritage.eu/ [20.02.2013].

Heritage Institutions of EU member states, and long running EU supported projects like 'Discovering the archaeologists of Europe'[11] – pan-European cooperation does not seem to be in bad shape at all.

Looking at it optimistically, this new level of collaboration may be a way to directly influence laws and regulations on national and EU level, and otherwise at least act as a channel for information about political decisions which may affect the "raw material" of the discipline or the conditions of employment. As a platform for research it can also work to mend the gaps between archaeology in different parts of Europe and make it easier to ask the big questions. However, there are several things to keep in mind here. With the creation of common guidelines, laws and codes of ethics, consolidation and the freezing of development on a restricted geopolitical level might also follow (see HAMILAKIS 2007). In the long term, chances of interpretations and practises becoming stuck on that level may also increase. As pointed out by archaeologist Matthew JOHNSON (2006), the basic ways in which archaeologists do things matter; the way we dig determines the way we interpret. Recent efforts to come to terms with – or at least understand – the differences between ways of doing archaeology in Europe, have provided a base for a deepened discussion (BIEHL *et al.* 2002; GRAMSCH – SOMMER 2011). But to actually harmonize practices and research frameworks on a European level, especially since it has been suggested that future archaeological frameworks should rely in part on financial support from the European Commission (HARDING 2009; KRISTIANSEN 2008; WILLEMS 1999; 2007), may prove a treacherous road to walk.

The question one might ask however, is if there not already such a thing as European archaeology. It was addressed in 2008 by Kristian Kristiansen and number of other scholars in a discussion article named: Do We Need the 'archaeology of Europe'? (KRISTIANSEN 2008). By demonstrating the narrow geographical scope of archaeological journals and their content, he argued that due to the step away from grand narratives and big questions, local and national presentations and restricted citation networks are dominating archaeological research. Subsequently, if we increase our knowledge of different languages and expand our field of enquiry, archaeology would regain its status as a scientific field of expertise. While calling attention to the relationship between ideology, politics and archaeology in the formation of critical knowledge, he argues that the best way to reach the global questions is through first establishing a European platform

[11] http://discovering-archaeologists.eu/ [12.05.2013].

(KRISTIANSEN 2008: 24). Yet, the distinction between a National, European and Global level is not very straightforward. We can talk of Europe in terms of east, west and central or expand it to 'the western world', but the opposition between nations within Europe and the differences in the self-image of nations as European, is not big enough to blur the contrast between Europe and its various Others. That line has been more actively maintained through the centuries. Using the words of Neal Ascherson in his response to the article, I am curious as to why Europe should somehow be the ultimate (or first) container (ASCHERSON 2008: 25). There seems to be far more reasons to avoid reinforcing anew the pre-set cognitive boundaries of our field. After all archaeology is one out of a number of western ways to tell stories about the human past, and while laws and practises differ there is in one sense already "no archaeology but European archaeology", as it was aptly put by Thomas Meier in his response (MEIER 2008: 36).

Narratives in the making

So what kinds of plots are suggested when projects apply to the Culture programme? And how might narratives, created in projects already considered to contain or to be able to generate European added value, actually read?[12] Based on previous discussions about the value of Europe in archaeology and the value of archaeology for the European Commission, a handful of very different archaeological projects funded by the Culture programmes will serve as examples of wider tendencies observed in my on-going research. First however, in order to get a better idea about the programmes themselves, I will briefly present some of the basics.

Between the years 2000 and 2006, Culture 2000 co-financed 1 529 actions and projects, out of which 398 were in the field of cultural heritage and with a total budget of 236,5 million Euro (2008: C3345/ECOTEC).[13] As you would expect, all of these projects did not involve archaeology or the care archaeological remains,

---

[12] In this part I use the definition of narrative as a story or a sequence within a larger story, presenting an idea or account of the past. Except for articles and books, this can refer to texts on a homepage, a project report (where the narrative of the project itself is told) or a folder/leaflet etc. Plot (not "emplotment") refers to aims or objectives, often with a meta-narrative backdrop (like for example EU policy texts which also have a narrative structure, especially memorandums).

[13] The number of projects is calculated based on the yearly lists of projects supported in the field of Cultural Heritage (http://ec.europa.eu/culture/archive/ [25.01.2013]).

but approximately a third had archaeological links or topics. Culture 2007–2013, still in force, has 400 million Euros for co-financing roughly 300 cultural actions per year.[14] Both programmes support cooperation activities among cultural operators from eligible countries and rest on the legal base first established in the Maastricht Treaty 1992.[15] Culture 2000 was meant to "share and highlight, at a European level, the common cultural heritage of European significance", and to promote "[…] mutual exchange between European and non-European cultures" (2000: 508/EC). Similarly, Culture 2007 is justified through phrases like:

> "For citizens to give their full support to, and participate fully in European integration, greater emphasis should be placed on their common cultural values and roots as a key element of their identity and their membership of a society founded on freedom, equity, democracy, respect for human dignity and integrity, tolerance and solidarity […]" (2006: 1855/EC).

The meaning of Europe is not addressed in these programmes, except when it comes to what is considered to be eligible 'Third countries' (geographically within Europe). There also remains a tension between an essentialist notion of unity, separating European and non-European cultures, and the respect for diversity (also reflected in the EU motto 'unity in diversity').

The programme is divided into three strands for different types of projects, but the one in focus here and the one providing the highest amounts of co-funding is strand number 1 (with subcategories 1.1 and 1.2.1), financing multiannual and shorter period cooperation projects.[16] The selection is carried out in three phases. The first takes place in the Commission where administrative eligibility criteria are checked. The second consists of a two-step evaluation by external experts in

---

[14] http://eacea.ec.europa.eu/culture/ [24.01.2013].

[15] Maastricht Treaty 1992, Article 128; Amsterdam Treaty 1997, Article 151; Lisbon Treaty 2009, Article 167.

[16] In the current programme, multi-annual projects have to involve a minimum of six cultural operators from six eligible countries. Lasting from a minimum of 36 months to a maximum of 60 months, they are granted from EUR 200.000 to a maximum EUR 500.000 per year. Shorter period cooperation projects should be shared by at least three cultural operators from three eligible countries. Maximum duration is 24 months and the amount of funding ranges between EUR 50.000 and EUR 200.000 per year. In both cases, EU support is limited to 50% of the total eligible costs (http://eacea.ec.europa.eu/culture/ [24.01.2013]).

which each application is assessed by two persons, separately at first and then together in a so called 'consensus meeting' where the final scores (1–100) are decided. The criteria of EAV, alike the second (relevance to the objectives of the Programme) and third criteria (level of excellence), counts for 0–20 points, while the remaining four can get only 0–10 points. In the third phase, the commission puts together a list of graded projects and makes a final decision based on the established budget, whereafter the expert evaluation sheets are sent out to the applicants together with a letter of rejection or acceptance. During the implementation of the project a commission 'project officer' is assigned as the main contact for the coordinators, and usually this is also the person evaluating the mid-term and final reports of the project (although the financial part is evaluated separately). In the final report, and in order to attain the last payment, the project is obliged to demonstrate how they achieved their goals and consequently also those of the sponsor.

European pasts

If we look at the archaeologies of a European past, there exist a range of projects corresponding to this in some way. In the short summary provided by the projects for the programme web page and database,[17] we find project descriptions such as:

> "The objective is to show the roots and the common European dimensions of comparable architecture or archaeological elements of the walled towns" (Culture 2000, Walled Towns: From Division to Co-Division, 2002–2003).

[17] While these elements, lifted from short descriptions, may provide an inadequate picture of the actual activities of these projects, it is the wider implications that are of concern here. Although it is clear that different types of project texts have different aims and therefore uses different types of language, this does not mean that texts directed towards the EU or the public are somehow less significant. Beside the structuring role of applications in organising projects, and the obligation to deliver something at least similar to what was promised, the mere fact that I am able to write this paper shows that these are lasting documents with an impact. Furthermore, it has been argued that the European institutions, due to their own ambivalence about what European culture and EAV actually means, have sought to conceptualise it using projects of an exemplary nature (BANUS 2002).

> "To search for a deeper knowledge of the influence that ancient Rome exerted on its provinces in order to help EU member states to be fully aware of the historical roots of their common cultural and social traditions" (Culture 2000, Living and Dying in the Roman Empire: New Perspectives of Funerary Archaeology, 2004–2005).

> "Archaeological experts will work on providing 3 cultural scenarios [for 3D reconstructions], based on Italian, German and Polish ancient societies, and on finding cultural interconnections, economic, cultural and artistic links between the scenarios" (Culture 2000, Connecting European Culture through New Technology – NetConnect, 2006–2009).

> "Cultural heritage offers tangible expression to the shared cultural past of European countries: it forms part of the 'source DNA' of the spirit that became the phenomenon Europe. Ultimately it represents the evidence and justification of Europe as a geo-cultural entity. Consequently, in the process of forging greater European unity, heritage institutions can play a significant role" (Culture 2007–2013, Cradles of European Culture, 2010–2015).[18]

There are three noticeable characteristics in these types of articulations; the search for similarities, the presumption that Europe can be treated as a culturally specific entity, and the meshing together of past and present, pointing towards Europe of today as the endpoint of a predestined development in line with the grand narratives of modernity. The language used, like 'European dimensions', 'roots' and 'common heritage/culture' echo EU policy. But does this also shine through in other project narratives? Taking a closer look at the NetConnect project, which created three virtual prehistoric Iron Age landscapes based on archaeological interpretation and research (Magna Graecia IT, Glauberg DE, Biskupin PL), the pattern is repeated is several places. One type of narrative produced was the visualizations themselves. While some of the reconstructions appear to be very site specific, they are also bound together in trailers created for the project, videos in which the 3D scenarios are put together like movie promotions with intense

[18] Collected from the lists of projects at: http://ec.europa.eu/culture/archive/culture2000/cult_2000_en.html [24.01.2013]; http://ec.europa.eu/culture/our-programmes-and-actions/projects-and-actions-supported_en.htm [24.01.2013].

music. Captions like 'The wait is over' and 'Nothing will be like before', binds the three sites together by displaying the most impressive features of the sites in sequence.[19] While these do not provide any insight as to what these sites might have in common, a paper from 2007 puts forward the following view:

> "At several times and in different regions, the impact of social, economic and environmental changes always caused new ways of living. The investigation of agriculture made people settle down and the emergence of new social systems sometimes caused the development of new settlement structures. Sharing the information about sites (connected with a common specification), spread all over Europe, links the past with the present and can give an idea of a future Europe without frontiers – frontiers in thinking and frontiers as borders and differentiation between countries" (BERTACCHINI *et al.* 2008: 6).

The meshing together of past and present is made extremely explicit in this paragraph, much more so than in other project publications (cf. BEUSING – POSLUSCHNY 2009). Not only is there an underlying presumption that sites all over prehistoric "Europe" can be easily connected, but also that highlighting these connections could assist in current integration of Europe as a geopolitical space. Even if there had been overwhelming empirical material to support a notion of a "pre-Europe", the issue lies in the circular argument. If projects, as in the case of Cradles of European Culture, start from the idea that cultural heritage can be used as "evidence and justification of Europe as a geo-cultural entity", the outcomes will have little or no chance of altering that statement. In the perspective of the European Commission, it can be argued that these types of plots and narratives fit their notion of a European identity and common heritage quite well, both in regards to the inherent and the cohesive values of cultural heritage. However, since Europe is no stable, sovereign, autonomous object but something that exists only in "historical relations and fields of power" (MALMBORG – STRÅTH 2002: 7), no two Europe's have been alike. This does not mean that Europe does not exist. When talking about the Europe of the EU as a construction and imagined community, it is important to remember that it is not just symbolic, but something

---

[19] [GT] NetConnect - Trailer 1. 2010. http://www.youtube.com/watch?v=DwYx5lj3wPk &feature=youtube_gdata_player [24.01.2013].

very real and important to those working within the European Commission; "they are 'building' Europe; they are 'constructing' a new world which they generally know to be right" (McDonald 1996: 47).

While I believe there can be ways for archaeologists to participate in such ventures, it requires some consideration as to what constructions "do". As pointed out by Bruno Latour, the more interesting question is instead whether a construction is well made or not (Latour 1999: 16). Looking at these first examples from the point of view of critical awareness, they are found wanting. After examining a number of other projects using this type of wording it is also clear that these passages often relate specifically to modules or activities meant to be presented to publics, and one might ask why such activities should require less critical consideration (instead of more) than the ones aimed towards the research community.

Europe as a "non"-value

For the second 'value' I will exemplify using only one project, partly because there are far fewer projects corresponding to this category and partly because this project makes for a very clear example. *Archives of European Archaeology* (AREA) began in 1999 as one of the European Commission's Cultural Heritage Laboratories and was funded in three intervals until 2008. As put forward in the European Commission summary in 2005, the project dealt with:

> "Issues such as the discovery and presentation of the European archaeological heritage, and the role it has played in the construction and consolidation of identities at regional, national and transnational levels".

In this context European archaeology is described as "a common cultural, scientific and socio-economic heritage of value and interest, as a source of knowledge and of leisure". The main activities consisted of research on the history of archaeology, and promotion and protection of archaeological archives.[20]

While the various descriptions on the webpage and the titles of narrative outputs contain words like 'pan-European', 'European archaeology', 'European

---

[20] http://ec.europa.eu/culture/archive/culture2000/pdf/projets2001/2001patrimoine_pluri.pdf [24.01.2013].

history' and 'European citizens',[21] this project does situate the discipline and the practises of archaeology in historical discourses of power, thereby elucidating how (as a social practice) it has come to affect the interpretations of the past. European archaeology becomes a European added value in itself, and perhaps this is the only sustainable way to approach discourses on European identity and a common heritage. After all, archaeology has been most valuable politically as a resource in building the Europe of nation states, a Europe with a dark legacy. AREA dealt with this legacy under themes like 'archaeology abroad' and 'archaeology under dictatorship'. In the travelling exhibition (2008) photos of archaeological excavations carried out under Franco and Mussolini were included,[22] and the project arranged meetings on topics like 'Networks, Contacts & Competition in the History of Archaeology' (2006). While it is hard to judge the ultimate impact of the project, it had the potential to act subversively while at the same time confirming Europe as a culturally situated space. The frame remains intact but the painting is altered dramatically.

When it comes to the added value for the EU, this project does perhaps not have a lot to offer in terms of identity building, but as with all larger cooperation projects funded by the culture programmes it involved archaeologists from many countries (11), and networks and platforms for cooperation were established as a result. In relation to such networks, in my interviews with project leaders and European Commission officials, the Europeanisation of archaeologists themselves has often come up. This brings us to what I have called a future archaeology of Europe. Perhaps these connections (and standardisation efforts) are the main result of all of these projects and if so, what might that mean for the future of the discipline?

A future archaeology of Europe?

Most projects co-funded by the Culture programmes have a strong network component. Mobility of professionals and the building of sustainable relations are some of their core features. Only the process of putting together an application creates such links, or as expressed by a European Commission official:

---

21 http://www.area-archives.org/ [24.01.2013].

22 http://www.area-archives.org/files/Pictures_AREA_Traveling_exhibition.pdf [24.01.2013].

> "[...] you work for I don't know how many months building a cooperation [...] if they are selected fine of course but even if they are not selected there is an advantage because the network is still there and can find other solutions" (EC-EAC01 2011, original language)".

But some projects are more focused on this aspect than others. One such project is 'Archaeolandscapes Europe' (ArcLand), running from 2010 to 2015. It is continuing the work of previous co-funded projects like 'European Landscapes – Past, Present and Future' (2004–2007) and 'Conservation through Aerial Archaeology' (2000–2001). The ultimate aim of the network is "the use throughout Europe of aerial survey and 'remote sensing' to promote understanding, conservation and public enjoyment of the shared landscape and archaeological heritage of the countries of the European Union", and the long term legacy is expected to be:

> "[...] better appreciation of the landscape and archaeological heritage of Europe, closer contact between heritage professionals and the general public, more effective conservation of the shared cultural heritage, the international sharing of skills and employment opportunities, better public and professional education, the wider use of archive resources and modern survey techniques, and higher professional standards in landscape exploration and conservation".[23]

While the word 'shared' is highlighted here (surely for political reasons), it appears to be referring to the countries of the EU in their current geopolitical form.[24] At the moment the project has 61 partners and associate partners from all EU member states, out of which 27 are co-organizers and tied to the EU contract.[25] With its focus on methodology development and networking, this project does

---

[23] http://eacea.ec.europa.eu/culture/funding/2009/selection/documents/strand_1_1/02_projects_description.pdf [24.01.2013].

[24] The synonym 'common' is more ambiguous in this sense, and more often refers to something we 'have in common'. We can 'share' something even if it is not something we have in common.

[25] ArchaeoLandscapes Europe, 1st Interim Report – Reference Period: 15/09/2011 to 31/12/2011 http://www.archaeolandscapes.eu/index.php/en/about/partners.html [24.01.2013].

indeed have the potential to increase the use and knowledge of archaeological practices in regards to aerial archaeology, and to change the level (literally and figuratively) of archaeological interpretation. Working across national borders is more or less prerequisite in this sense, and the outcomes can create a better base to ask the larger questions as called for by KRISTIANSEN (2008). Nevertheless, it can also consolidate "the landscape and archaeological heritage of Europe" in its singular form, making it much easier to ask questions about – or find patterns within – prehistoric "Europe", while turning "non-European" landscapes into blind spots (see also MEIER 2013 on the "Europeanness" of landscapes).

Lastly, a second project worth mentioning that belong to this category is 'Archaeology in Contemporary Europe – ACE' (2007–2012). It builds on a network of 13 partners and focuses on archaeology as a cultural, scientific and economic practice in the present. Similarly to the AREA project, it deals with archaeology as a European added value in itself, with themes like: the significance of the past in the present, current practices in archaeology, mapping of archaeological practitioners and increasing the relevance and public interest of archaeology.[26] The importance of questions on the nature of archaeology in today's Europe cannot be understated, but it needs to be accompanied by critical questions like: What is Europe about in today's archaeology? Alongside this contemporary focus and interest in building a base which could be used for streamlining measures, it is interesting to note that they are also filling the frame of Europe with archaeological interpretations. Through narrative compilations like 'The very beginning of Europe?: cultural and social dimensions of Early-Medieval migration and colonisation' (ANNAERT *et al.* 2012), and 'L'Europe: Un continent redécouvert par l'archéologie' (DEMOULE 2009), they engage with the question of a European past (critically in parts), but even so they end up confirming the specificity of the continent.

Conclusions

> "No other world region holds such a rich inheritance, reflecting the way in which Europe has been the cradle of so much of the world's development. But we are still not reaping the full benefits and spill-over effects of culture in Europe" (EU Commissioner in culture Androulla VASSILIOU, SPEECH/12/725, original language).

26 http://www.ace-archaeology.eu/ [24.01.2013].

Archaeology of Europe, European archaeology, or looking for a European 'unity in diversity' in the remains of the past – does it really matter which way we turn it? It certainly does, but it would be naïve to think that they are mutually exclusive. The adjective 'European' is still used extensively in archaeology. Talking about European archaeology seems closer at hand than talking about archaeology in Europe and reading for instance the session descriptions for *European* Association of Archaeologists annual meeting in Pilzen 2013, interpretations of periods in *European* prehistory, such as the *European* Neolithic, or certain pan-*European* phenomenon's, are addressed from a *European* perspective.[27] In such forms it is generally meant to indicate the current geographical extent of the continent carrying the name, although being simultaneously a contemporary signifier and a construct (or several) – relating to historically, socially, and institutionally specific structures of beliefs and statements (FOUCAULT 1971) – calling something European comes with added baggage.

When archaeologist Colin Renfrew wrote about Europe and archaeology in the 1990s, he insisted that we need to re-define the critical questions and "[...] formulate them in relation to frameworks of thought that do not themselves offer or determine a readymade answer" (RENFREW 1994: 171). Despite this, the argument still rested on the presumption that "[...] every continent is unique" (RENFREW 1994: 157). While it is true that "[...] we can only change the structures that bind us once they have been thought" (HODDER 1990: 19), it cannot stop there. Once we accept Europe as something constructed, as both signifier and discourse of power, it is not enough to just recognize this. At that moment it can no longer serve as a valid base or a frame for asking questions about societies preceding its conception by millennia. As insisted by David Lowenthal "[...] any treatment of the past, however circumspect, invariably alters it" (LOWENTHAL 1985: 264), and I would like add that any change in the formations of disciplines, any new or revamped frames we set around our topics, has bearing on our interpretations. Buzzwords and catchphrases, adapted to a pre-set format in order to maximize chances of achieving funding, may seem like external circumstances having nothing to do with the actual activities carried out. But only by working in a larger geopolitical frame and being encouraged to think through concepts such as European added value we engage with them.

In this paper I have discussed a number of ways in which the first application criteria of the European Commission Culture programmes correlate to, or is

[27] http://www.eaa2013.cz/programme/ [24.04.2013].

manifested in, European Commission discourse and in narratives of archaeology and heritage, pointing towards a number of potential effects. From this discussion it is clear that the notion of EAV has a critical and subversive potential, while it could also lead to the consolidation of ethics and platforms for cooperation on EU level. In some cases the recycling of longstanding notions of European-ness become visible. In the late 1990s, archaeologist Mark Pluciennik wrote in response to the increased prospects of European funding that, as collaborating archaeologists

> "[...] we should consciously, as part of our task, work deliberately on the margins of any proposed 'European' boundaries",

and that

> "[...] projects which deliberately ignore physical and political boundaries by working across them may have more to offer in this respect than either collaborative projects within individual countries or pan-European projects aimed at producing common methodologies, standards and archaeological languages" (PLUCIENNIK 1998: 822).

Fourteen years later, I like to consider my research as a way to revisit this request. Not because increased cross-border cooperation, or the European Union for that matter, should somehow be seen as destructive forces or as having a negative effect on archaeologies in Europe and elsewhere. Not because there are no warranted empirically oriented pan-European research projects, or because grand narratives of Europe are somehow fictional accounts, but because projects taking EU integration goals a point of origin and Europe as a contrast frame for their research, should also critically reflect on the foundations of said work. There may be no Archaeology but European archaeology, but it surely has more values than European added value. In relation to Council of Europe's European landscape convention, Thomas Meier has discussed the potential of viewing the European-ness of landscapes as the manner and values by which they are managed (freedom, democracy etc.) in order to work against traditional exclusive and top-down approaches (MEIER 2013). This might be one way to tackle the issue, even though it does not challenge the frame. These values are embedded in the dominant western meta-narratives and as such intimately related to "[...] the impulse to moralize reality, that is, to identify it with the social system that is the source of any morality that we can imagine" (WHITE 1987: 14). Also, as we know from postcolonial discourses they can be exclusive, albeit in different ways.

Lastly, in light of current European Commission initiatives, it appears as if archaeological activities and heritage processes are less interesting as objects of study or areas of research, than they are as a selling point. And, that where there is an anxiety about the usefulness of heritage in relation to the idea of Europe, rather than being reconsidered, essentialist ideas of Europe are being reused. Archaeologist Cornelius Holtorf has argued that "archaeology matters most when its meta-stories matter" (HOLTORF 2010: 391), highlighting the point that we need to continue to tell stories but become better at it, by understanding and critically assessing the meta-stories we tell and the ones we evoke. Archaeologists may never be able to achieve genuine dialogue with their narrative subjects; but in reflecting upon their situated authorship they might be able to do so with the people upholding the structures enabling their narratives to be written.

References

ADDELSON, KATHRYN PYNE (2003 [1983]), The man of professional wisdom. In: Sandra Harding – Merrill Hintikka (eds), *Discovering reality: Feminist perspectives on epistemology, metaphysics, methodology, and philosophy of science*. Dordrecht. 165–186.

AITCHISON, KENNETH (2000), The funding of professional archaeological practice in England. *Cultural Trends* 10/39, 1–32.

AITCHISON, KENNETH (2009), After the 'gold rush': global archaeology in 2009. *World Archaeology* 41/4, 659–671.

AITCHISON, KENNETH – NATHAN SCHLANGER (eds) (2010), *Archaeology and the global economic crisis: Multiple impacts, possible solutions*. Brussels.

ANNAERT, RICA – TINNE JACOBS – INGRID IN'T VEN (eds) (2012), *The very beginning of Europe?: cultural and social dimensions of Early-Medieval migration and colonisation (5th–8th century)*. Conference proceedings: Archaeology in contemporary Europe, May 17–19 2011. Brussels.

ASCHERSON, NEAL (2008), Europe and its "Kossinna province." *Archaeological Dialogues* 15/Special Issue 01, 25–27.

BALZARETTI, ROSSI (1992), The creation of Europe*. *History Workshop Journal* 33, 181–196.

Banús, Enrique (2002), Cultural policy in the EU and the European identity. In: Mary Farrell – Stefano Fella – Michael Newman (eds), *European integration in the 21st century: unity in diversity?* London. 158–184.

Barroso, José Manuel Durão (2012), *Opening of 'Europe Meets the World' Exhibition.* Copenhagen, 12 January 2012. SPEECH/12/7. (http://europa.eu/rapid/press-release_SPEECH-12-7_en.htm [24.07.2013]).

Bernal, Martin (1991), *Black Athena: the Afroasiatic roots of classical civilization. Vol.1, The fabrication of Ancient Greece 1785–1985.* London.

Bertacchini, Pier Augusto – Giuseppe Conti – Ruth Beusing – Raffaele De Amicis– Aleksander Bursche – Markus Etz – Daniel Holweg – Axel Posluschny – Assunta Tavernise – Maria T. Linaza – Douglas Pritchard – Thomas W. Maver – Susanne Sievers (2008), NetConnect: Connecting European culture through new technology. *Conference proceedings: EVA-London 2007.* London.

Beusing, Ruth – Axel Posluschny (2009), The NetConnect-Project – New access to ancient sites. *Conference proceedings: Computer Applications to Archaeology* 2009 Williamsburg, USA. (http://www.caa2009.org/PapersProceedings.cfm [1]).

Biehl, Peter F. – Alexander Gramsch – Arkadiusz Marciniak (eds) (2002), *Archaeologies of Europe. History, methods and theories/ Archäologien Europas. Geschichte, Methoden Und Theorien.* Tübinger Archäologische Taschenbücher 3. Münster – New York – München – Berlin.

Bugge, Peter (2003), A European cultural heritage? Reflections on a concept and a programme. In: Robert S. Peckham (ed.), *Rethinking heritage; cultures and politics in Europe.* London – New York. 1–13.

Burke, Peter (1980), Did Europe exist before 1700? *History of European Ideas* 1, 21–29.

Burke, Peter (2006), How to write a history of Europe: Europe, Europes, Eurasia. *European Review* 14/02, 233–239.

Childe, Vere Gordon (1925), *The dawn of European civilization.* The history of civilization series. London.

Childe, Vere Gordon (1930), *The Bronze Age.* Cambridge.

Childe, Vere Gordon (1962), *The prehistory of Europe.* London.

Clark, Grahame (1952), *Prehistoric Europe: The economic basis.* London.

Cleere, Henry (2001), The uneasy bedfellows: Universality and cultural heritage. In: Robert Layton – Peter G. Stone – Julian Thomas (eds), *Destruction and conservation of cultural property*. London. 22–29.

Council of Europe (1994), *The Bronze Age – The first golden age of Europe*. European Heritage 2. Strasbourg.

Delanty, Gerard (1995), *Inventing Europe: idea, identity, reality*. New York.

Delanty, Gerard (2010), *The European heritage from a critical cosmopolitan perspective*. LSE "Europe in Question" Discussion Paper Series 19. London.

Demoule, Jean-Paul (2004), Archéologie préventive, recherché scientifique et concurrence commercial. In: Pierre-Laurent Frier (ed.), *Le nouveau droit de l'archéologie préventive*. Paris. 199–242.

Demoule, Jean-Paul (ed.) (2009), *L'Europe: un continent redécouvert par l'archéologie*. Paris.

Demoule, Jean-Paul (2011), We still have to excavate – but not at any price. *Archaeological Dialogues* 18/01, 5–10.

Díaz-Andreu García, Margarita (2007), *A world history of nineteenth-century archaeology: Nationalism, colonialism, and the past*. Oxford.

During, Roel (2010),. *Cultural heritage discourses and Europeanisation: discursive embedding of cultural heritage in Europe of the regions*. Wageningen.

English, Penny (2008), "The flowering of the cultures": The European Union and regional identity. In: Francis P. McManamon – Andrew Stout – Jodi A. Barnes (eds), *Managing archaeological resources: Global context, national programs, local actions*. Walnut Creek. 58–67.

Fairclough, Norman (1992), *Discourse and social change*. Cambridge.

Fairclough, Norman (1995), *Critical discourse analysis: the critical study of language*. Language in Social Life Series. London.

Foucault, Michel (1971), *L'ordre du discours: leçon inaugurale au Collège de France prononcée le 2 décembre 1970*. Paris.

Goddard, Victoria A. – Josep R. Llobera – Cris Shore (1994), *The anthropology of Europe: Identity and boundaries in conflict*. Explorations in Anthropology. Oxford.

Graeber, David (2001), *Toward an anthropological theory of value: the false coin of our own dreams*. New York.

GRAMSCH, ALEXANDER (2000), "Reflexiveness" in archaeology, nationalism, and Europeanism. *Archaeological Dialogues* 7, 4–19.

GRAMSCH, ALEXANDER – ULRIKE SOMMER (eds) (2011), *A history of central European archaeology: Theory, methods, and politics*. Budapest.

GRAVES-BROWN, PAUL – SIAN JONES (eds) (1996), *Cultural identity and archaeology: The construction of European communities*. London.

GRÖHN, ANNA (2004), *Positioning the Bronze Age in social theory and research context*. Acta Archaeologica Lundensia Series 8/47. Stockholm.

HAMILAKIS, YANNIS (1999), La trahison des archeologues? Archaeological practice as intellectual activity in postmodernity. *Journal of Mediterranean Archaeology* 12, 60–79.

HAMILAKIS, YANNIS (2007), From ethics to politics. Introduction to Yannis Hamilakis – Philip Duke (eds), *Archaeology and capitalism: From ethics to politics*. Walnut Creek. 15–40.

HARDING, ANTHONY (2009), Towards a European archaeology. *World Archaeology* 41/4, 629–640.

HAWKES, C. F. CHRISTOPHER (1940), *The prehistoric foundations of Europe to the Mycenean Age*. London.

HØLLELAND, HERDIS (2008), *The Bronze Age – the dawn of European civilisation?: A case study of usages of the past within the European identity discourse*. Master Thesis, University of Oslo. Oslo.

HØLLELAND, HERDIS (2010), Spells of history: Childe's contribution to the European identity discourse. *Bulletin of the History of Archaeology* 20/1, 30–37.

HØLLELAND, HERDIS (2011), Personhoods for Europe: The archaeological construction and deconstruction of European-ness. In: Christopher Prescott – Håkon Glørstad (eds), *Becoming European: The transformation of third millennium Northern and Western Europe*. Oxford. 12–18.

HODDER, IAN (1990), *The domestication of Europe: Structure and contingency in Neolithic societies*. Oxford.

HOLTORF, CORNELIUS (2010), Meta-stories of archaeology. *World Archaeology* 42/3, 381–393.

JOHNSON, MATTHEW H (2006), On the nature of theoretical archaeology and archaeological theory. *Archaeological Dialogues* 13/02, 117–132.

KOHL, PHILIP L. – CLARE FAWCETT (1996), *Nationalism, politics and the practice of archaeology*. Cambridge.

KRISTIANSEN, KRISTIAN (2008), Do we need the 'archaeology of Europe'? *Archaeological Dialogues* 15/Special Issue 01, 5–25.

LAMONT, MICHÉLE (2009), *How professors think: Inside the curious world of academic judgment*. Cambridge.

LATOUR, BRUNO (1999), *Pandora's hope: essays on the reality of science studies*. Cambridge.

LOWENTHAL, DAVID (1985), *The past is a foreign country*. Cambridge.

LYOTARD, JEAN-FRANÇOIS (1984), *The postmodern condition: a report on knowledge*. Theory and History of Literature. Manchester.

MALMBORG, MIKAEL – BO STRÅTH (2002), *The meaning of Europe: variety and contention within and among nations*. Oxford.

MCDONALD, MARYON (1996), 'Unity in diversity': Some tensions in the construction of Europe. *Social Anthropology* 4, 47–60.

MEIER, THOMAS (2008), There is no archaeology but European archaeology … *Archaeological Dialogues* 15/Special Issue 01, 34–36.

MEIER, THOMAS (2013), What's European in the landscape? In: Dirk Callebaut – Jan Mařík – Jana Maříková-Kubková (eds), *Heritage reinvents Europe*. Proceedings of the International Conference Ename, Belgium, 17–19 March 2011. Budapest. 55–63.

NIKLASSON, ELISABETH (2013), And we would like to thank: The role of funding in archaeology. In: Anna Källén (ed.), *Making cultural history: new perspectives on Western heritage*. Lund. 151–161.

PAGDEN, ANTHONY (2002), *The idea of Europe: from antiquity to the European Union*. Washington.

PETROVIC, TANJA (2011), Thinking Europe without thinking: Neo-colonial discourse on and in the western Balkans. *Eurozine* 2011-09-22. (http://www.eurozine.com/articles/2013-02-27-petrovic-it.html [24.01.2013]).

PLUCIENNIK, MARK (1998), Archaeology, archaeologists and 'Europe'. *Antiquity* 72/278, 816–824.

PLUCIENNIK, MARK (1999), Archaeological narratives and other ways of telling. *Current Anthropology* 40/5, 653–678.

RENFREW, COLIN (1973), *Before civilization, the radiocarbon revolution and prehistoric Europe.* London.

RENFREW, COLIN (1994), The identity of Europe in prehistoric archaeology. *Journal of European Archaeology* 2/2, 153–174.

RIEKMANN, MONICA (1997), The myth of European unity. In: Geoffrey Hosking – George Schöpflin (eds), *Myths and nationhood.* New York. 60–71.

ROSAMOND, BEN (2000), *Theories of European integration.* The European Union series. Basingstoke.

SAID, EDWARD W. (1978), *Orientalism.* London.

SCHUBACK, MARCÍA S. C. (2013), *Traditionens hermeneutik* [The hermeneutics of tradition]. Lecture/Presentation 15 February 2013, Stockholm University, unpublished.

SHANKS, MICHAEL (2004), Archaeology and politics. In: John L. Bintliff – Timothy K. Earle – Christopher S. Peebles (eds), *A companion to archaeology.* Malden. 490–508.

SHORE, CRIS (2000), *Building Europe: The cultural politics of European integration.* London.

SHORE, CRIS (2006), 'In Uno Plures'(?) EU cultural policy and the governance of Europe. *Cultural Analysis* 5, 7–26.

SMITH, LAURA-JANE (2006), *Uses of heritage.* Abingdon.

SPIVAK, GAYATRI CHAKRAVORTY (1988), Can the subaltern speak? In: Cary Nelson – Lawrence Grossberg (eds), *Marxism and the interpretation of culture.* Illinois. 271–313.

TARSCHYS, DANIEL (2005), *The enigma of European added value: setting priorities for the European Union.* SIEPS 4. Stockholm.

THEILER, TOBIAS (2006), *Political symbolism and European integration.* Manchester.

THOMAS, JULIAN (2004), *Archaeology and modernity.* London.

TRETTER, ELIOT (2011), The 'value' of Europe: The political economy of culture in the European Community. *Geopolitics* 16/4, 926–948.

TRIGGER, BRUCE G. (2006 [1989]), *A history of archaeological thought.* Cambridge.

TZANIDAKI, JOHANNA D. (2000), Rome, Maastricht and Amsterdam. *Archaeological Dialogues* 7/01, 20–33.

VASSILIOU, ANDROULLA (2012), *Celebrating Europe's extraordinary cultural heritage*. Published speech from the 5th European Heritage Days Forum in Cyprus Nicosia, 15 October 2012. SPEECH/12/725. (http://europa.eu/rapid/press-release_SPEECH-12-725_en.htm [24.01.2013]).

VOS, CLASKE (2011), Negotiating Serbia's Europeanness: on the formation and appropriation of European heritage policy in Serbia. *History and Anthropology* 22/2, 221–242.

WHITE, HAYDEN (1987), *The content of the form: narrative discourse and historical representation*. Baltimore.

WILLEMS, WILLEM J. H. (1999), *The future of European archaeology*. Oxbow Lectures 3. (https://openaccess.leidenuniv.nl/bitstream/handle/1887/9860/1_953_003.pdf?sequence=1 [24.01.2013]).

WILLEMS, WILLEM J. H. (2007), The work of making Malta: The Council of Europe's Archaeology and Planning Committee 1988–1996. *European Journal of Archaeology* 10/1, 57–71.

WILLEMS, WILLEM J. H. (2009), Archaeological resource management and academic archaeology in Europe: Some observations. In: Silvia Alaura – Anna Lucia D'Agata (eds), *Quale futuro per l'archeologia*. Roma. 89–99.

EU and CoE Documents

(1957) *Treaty establishing the European Economic Community*. Treaty of Rome.

(1973) *Declaration on the European Identity*. The Copenhagen Summit, 14–15 December. Bulletin of the European Communities 12/73, 118–122.

(1974) European Commission (EEC). *Commission Recommendation of 20 December 1974 to Member States concerning the protection of the architectural and natural heritage* (75/65/EEC). OJ L 021 (28/01/1975), 22–23.

(1977) European Commission (EEC). *Community action in the cultural sector Communication to the Council*, COM (77) 560 final, 2 December 1977. Bulletin of the European Communities supp. 6/77.

(1981) Council of the European Communities (CEC). *Memorandum on Revitalization of the Community*, presented by the French Government. Bulletin of the European Communities 14/11/1981, point 3.5.1, I.

(1982) European Parliament (EP). *Debate concerning 'EP 1982/0206 – Report on the protection of the architectural and archaeological heritage'*. Debates of the European Parliament. Annex to OJ, 1-288 (13/09/82), 7–18.

(1992) *Treaty on European Union*. OJ C 191/01 (27/9/1992). Treaty of Maastricht.

(1993) Council of Europe (CoE), *The Bronze Age: the first golden age of Europe* (Pamphlet). Cultural Heritage Division, Strasbourg.

(1996) Council of Europe (CoE). *European cultural co-operation: activities of the European Union and relations with the Council of Europe*. Parliamentary Assembly 7th of June 1996. Doc. 7575.

(1997) European Parliament and the Council. *Decision No. 2228/97/EC of the European Parliament and of the Council of 13 October 1997 establishing a Community action programme in the field of cultural heritage (Raphael)*, OJ L 305 (8/11/1997).

(2000) European Parliament and the Council. *Decision No. 508/2000/EC of the European Parliament and of the Council of 14 February 2000 establishing the Culture 2000 programme*. OJ L 63/1 ((10/3/2000).

(2000–2006) European Commission. Culture 2000. Lists of *Projects supported in the field of Cultural Heritage*, divided per year. (http://ec.europa.eu/culture/archive/culture2000/project_annuel/projects1_en.html [24.01.2013]).

(2003) Council of the European Union. *Council Resolution of 19 December 2002 implementing the work plan on European cooperation in the field of culture: European added value and mobility of persons and circulation of works in the cultural sector.* OJ C 013 (18/01/2003).

(2003) European Commission. Call for Proposals: *A cultural celebration of the enlargement of the European Union.* Brussels, 12 September 2003. (http://www.eurodesk.no/element_db/4/4_Cultural_Celebration_of_EU_Enlargement.pdf [24.01.2013]).

(2005) UNESCO. *Convention on the protection and promotion of the diversity of cultural expressions*. Paris 20 October 2005. (http://unesdoc.unesco.org/images/0014/001429/142919e.pdf [24.01.2013]).

(2006) European Parliament and the Council. *Decision No 1855/2006/EC of the European Parliament and of the Council of 12 December 2006 establishing the Culture Programme (2007 to 2013).* OJ L 372/1 (27/12/2006).

(2007–2013) European Commission. Lists of *Projects supported in the field of Cultural Heritage*. (http://ec.europa.eu/culture/our-programmes-and-actions/projects-and-actions-supported_en.htm [24.01.2013]).

(2008) European Commission. *Final External Evaluation of the Culture 2000 programme (2000–2006)*. C3345. Framework Contract on Evaluation, Impact Assessment and Related Services. ECOTEC. United Kingdom.

(2010) European Commission. *Programme Guide, Culture 2007–2013*. Education, Audiovisual and Culture Executive Agency. (http://ec.europa.eu/culture/calls-for-proposals/culture-programme-guide_en.htm [24.01.2013]).

(2011) European Parliament and the Council. *Decision No 1194/2011/EU of the European Parliament and of the Council of 16 November 2011 establishing a European Union action for the European Heritage Label*. OJ L 303/1 (22.11.2011).

Interviews

EC-EAC01/2011, European Commission official – Culture Programme Management, Brussels November 2011.

EC-EAC03/2012, European Commission official – Commissioners Cabinet Culture, Brussels November 2012.

Archaeological nature writing in the making of past landscapes – an ecocritical approach to the use of wilderness narratives in Finnish archaeology

TUIJA KIRKINEN

Abstract:
In this paper the genre of archaeological writing referred to as 'nature writing' will be discussed in relation to the archaeological grand narratives of Finland. Special attention will be placed on the idea of wilderness due to its central role in Finnish environmental thinking, as well as in Finnish archaeology, both as a source of welfare and as the origins of Finnish culture. By using an ecocritical approach, an interdisciplinary method for analysing literary and scientific texts as cultural constructions of nature, a selection of Finnish prehistory books and scientific papers will be discussed in relation to their impact on constructions of 'a past in nature' and in archaeological interpretations. Based on the results of this discussion, I argue that the circulation of international environmental narratives in archaeological texts can be discerned, as can the influence of Judeo-Christian tradition. Furthermore, it shows the impact of the literal wilderness genre and the Georgic tradition together with international waves of nationalism, national romanticism and environmentalism. The recognition of these legacies inherent in Finnish narratives is elementary for many reasons, most of all when it comes to identifying the political influences of the wilderness narrative.

Introduction

> "In terms of the wilderness, Finland is a European superpower. Its forest resources are exceeded only by Russia and Sweden, and the number of its lakes, about 188,000, is second to none [...] As a result, there are good reasons to suppose that in these circumstances, the environment must have played a central role in Finnish history" (MYLLYNTAUS 2002).

When discussing the special role of forests in Finnish society, economy, environmental politics and national identity, it is important to remember that one of the main origins can be found in Finland's location, at the edge of the Taiga on the northern limits of cultivated Europe. Just to get an idea of the Finnish environment as seen in numbers, forests cover about 85 % of the land area (FINNISH FOREST ASSOCIATION 2012). With a sparse population to begin with, about 70 % of Finns live in an area that covers only 5 % of the land area (FINNISH ENVIRONMENT INSTITUTE 2013). However, taking a step back, I will now ask you to re-think the above mentioned classifications of forests versus inhabited areas as it carries with it the legacies of dualistic divisions between nature and culture, the very legacies of concern in this paper. This dualism has formed the basis for two of the most influential narratives about nature in the western world, of wilderness and pastoral lands, the former usually seen as representing the myth of original nature (*Fig. 1*).

Fig. 1. An Eastern Finnish view from Koli towards Lake Pielinen is the most iconographic representation of Finnish wilderness, visited by voyagers from the 19th century onwards and memorialized by artists and photographers (Photo T. Kirkinen).

Ever since its creation as a discipline, the notion of wilderness has had a strong explanatory power in Finnish archaeology, both when it comes to the origins of Finnish culture and matters like welfare and subsistence (ASPELIN 1880; 1885; TALLGREN 1931: 215–223; KIVIKOSKI 1961: 277–278). Due to its importance and longevity, the focus of this text is set already on the early narratives and the timespan between the 1870s and 1930s, even if later texts will also be discussed. At this time western wilderness narratives were adopted and formatted into Finnish archaeology. As I will show, this was done mainly in order to explain the past in archaeologically low-intensive areas which then in turn led to the construction of the idea of wilderness as past. Moreover, the political nature and impact of these narratives will be addressed by taking the role of the narrator and the concept of 'otherness' into consideration. As a final point, the concrete, long term influences of wilderness narratives on Finnish archaeology will be discussed, pointing to four different ways in which they have figured in archaeological interpretation; wilderness as an uncolonized area; wilderness as pristine; wilderness as wild and wilderness as a stage for male culture.

The roots of the western wilderness narrative,[1] founded on the idea of nature as uncontaminated by civilization, have since long been detected in Judeo-Christian literature, updated in pace with modernisation through the ideas of environmentalism, romanticism and nationalism (CRONON 1996; GARRARD 2004: 59–84; OELSCHLAEGER 1991; SIMMONS 1993: 14–15). In archaeology, the concept of wilderness has generally referred to a special state in the history of mankind, the one of nature before humans or nature without humans, registered as an archaeologically empty area. Although in some cases it has also been used in archaeology to describe sparsely populated areas, especially in relation to the hunter-gatherer mode of living in harmony with nature (OELSCHLAEGER 1991: 173–174). However, a change in the approach to wilderness occurred in the first decade of the 21st century, when several international interdisciplinary projects between archaeologists and ecologists was initiated, the aim of which was to study the land-use legacies in areas formerly thought of as pristine or not affected by humans (BRIGGS *et al.* 2006; FOSTER *et al.* 2003; HAYASHIDA 2005).

---

1 The word 'wilderness' is thought to derive from the Bible translations into English, stemming from the Anglo-Saxon word 'wilddeoren', which means wild beasts living outside the cultivated areas (GARRARD 2004: 60). The word 'wild' might also derive from the old English expression 'weald' or 'waeld', meaning a forest (OELSCHLAEGER 1991: 356–357; see also HARRISON 1992: 201).

In Finnish archaeology, the term 'erämaa', literally meaning 'hunting ground', primarily refers to the historically known tradition of wilderness-use, in which the forests were used by neighbouring farmers for fishing and fur hunting (NORDMAN 1924: 188–189; TALLGREN 1931: 215–223; TAAVITSAINEN *et al.* 1998: 236–237). As such, wilderness was constructed in reference to the farming population. This way of referring to wilderness has parallels to the archaeological idea, that the division between nature and culture has its origins in the processes of neolithization as the beginning of man's active role in environmental change through domestication (OELSCHLAEGER 1991: 1–30).

In the following, before going into the concept of wilderness as used in Finnish archaeology, I will try to give a better idea about the selected texts and the ecocritical approach itself, which focuses on various expressions of nature.

Applying ecocritical reading to Finnish archaeology

Ecocriticism (aslo called green cultural theory or environmental criticism) can be defined as a field of study which takes an environmental perspective as its point of origin in the analysis of literal and cultural products. The roots of ecocriticism can be traced back to the modern environmental consciousness and the rise of environmentally oriented academic fields during the 1970s, but it was not until the 1990s that it became more mainstreamed and was used actively within a number of disciplines. As an umbrella term, ecocriticism accommodates a very theoretically and methodologically diverse field that focuses on various expressions of nature, both within specific historical and cultural frames, and inside different literal and aesthetical genres (BUELL 2005: 2–13, 138; GARRARD 2004:1–14; GLOTFELTY 1996). With the study of environmental narratives as its main topic of study, ecocriticism often overlaps with narrative theory, although the dialogue between these two well-established approaches has been vague thus far.

Although environmental narratives are most familiar to us as presented in fiction and in popular culture, they can also be identified in scientific writings as underlying constructions of nature (BUELL 2005: viii, 17–21, 138; HOWARTH 1996: 74–75; LOVE 1999; 2003: 37–64). For example, the apocalyptic rhetoric originating from Revelation, the last book of the Bible, can be found reproduced in scientific discourses about the threats of global warming. Also, the use of romantic pastoral perspectives and the accompanying ideas of balance and harmony in nature have had an impact on ecological discourse (GARRARD 2004:

5–10, 56–58, 93–107; HEISE 1999; HOWARTH 1996: 74–75; LOVE 1999). As a genre within the study of environmental narratives, wilderness has been a central topic for ecocritical readings from the 1990s onwards (BUELL 1995; GARRARD 2004: 59–84; LOVE 2003: 85–88; see also CRONON 1996; OELSCHLAEGER 1991).

The texts selected for ecocritical reading in this study consisted of a collection of Finnish prehistory books, texts on local prehistories and a number of papers, dating from the 1870s up to the present day. While not covering everything possible, the research material does include all major Finnish Prehistory books written by professional archaeologists. The reasoning behind this choice was that, compared to research articles addressed to academic readers, the prehistory books often employ classic narrative forms and were commonly used to promote archaeology and popularize prehistory for the wider public (see JOYCE *et al.* 2002: 121). The coherent structure of the narrative forms also proved to be useful in interpreting archaeological finds, to construct a coherent spatiotemporal whole. It was especially useful when it came to 'filling the blanks' in areas which lacked archaeological finds. On a deeper level, these national prehistories drew heavily on the cultural-historical core narrative of the colonizing of Finland, forming a part of Finnish nationalistic history writing. As have been noted before, this mode of narratives commonly used the idea of wilderness in interpretations of the colonization of archaeologically low intensive areas. Finally, the wide circulation of these books, used for example as university textbooks in educating new generations of archaeologists, underlines their influence on Finnish archaeology.

The basis for this genre of popularised archaeological narratives was laid in 1875 by Johannes Reinhold Aspelin (1842–1915) (ASPELIN 1875) – also called "the father of Finnish archaeology" – when his doctoral thesis *Suomalais-ugrilaisen muinaistutkinnon alkeita* (The basics of Fenno-Ugrian archaeology) was published. A few years later, in 1880, he added the thesis *Muinaisjäännöksiä Suomen suvun asumus-aloilta* (Ancient monuments from the neighbourhood of Finnish folk) to this body of work, and in 1885 the next thesis *Suomen asukkaat pakanuuden aikana* (Finland's inhabitants during the heathen era) was published. Aspelin was the first academically educated archaeologist in Finland, working in the contemporary Grand Duchy of Finland (1809–1917) under the rule of the Russian Empire. Naturally his works have to be seen in their historical context, since Aspelin together with other academics of this time (like ethnologists and folklorists), considered his work to be part of a strong nationalistic mission in which establishing the origins of the Finnish nation would justify it politically as a nation state (HAILA 2003: 179–180; MIKKELI 1992; SIMOLA 2008a).

After Aspelin, the next grand narratives of Finnish prehistory was authored by Aarne Michaël TALLGREN (1931), Ella KIVIKOSKI (1961), Matti HUURRE (1979), Unto SALO (1984) and Pirkko-Liisa LEHTOSALO-HILANDER (1984), Ari SIIRIÄINEN (1987) and Torsten EDGREN (1993). Most of the texts were written in Finnish, with the exception of EDGREN (1993) who wrote in Swedish, and KIVIKOSKI (1961) whose book was published in both Swedish (1964) and in English (1967). Additionally, in the course of writing this article, I found out that a new book on prehistory in Finland will be published in the coming year, but the manuscript is unfortunately not yet available and could therefore not be included in the analysis. Besides Prehistory books, and in order to complement the narratives stretching over 130 years of Finnish archaeology, a sample of local prehistories, survey reports and scientific articles concentrating on the prehistory of wilderness areas were selected. Despite this, the list is by no means complete but should be seen as a sample of what forms Finnish wilderness writing may take in Finnish prehistory books.

Judeao-Christian tradition in archaeological wilderness narratives

Through the works of Aspelin, the concept of wilderness was introduced to Finnish archaeology at the very beginning of the discipline. By collecting and reviewing the existing information about archaeological sites, Aspelin produced a geographically nuanced vision of Finland with its archaeologically rich south-western parts and eastern Karelia, as well as the low intensive areas in the northern and eastern parts of Finland (*Fig. 2*). In order to explain the past of these white or "empty" areas, Aspelin referred to them as wilderness regions inhabited by non-Finnish hunter-gatherer populations, used by neighbouring farmers for hunting. In his works, the state of prehistoric wilderness is portrayed as a specific stage which ended with the increased clearance of forests by Finns. By doing so he provided the foundations for how the idea of wilderness and its role in the colonization of Finland, would be used in Finnish archaeology for decades to come (ASPELIN 1880: 250, 261; 1885: 30, 62–63, 80–82).

Scientifically, Aspelin based his theory of prehistoric wilderness areas on the historically known use of forested areas, as well as the land-owning tradition developed to organize the use of forest resources among farmers (PAASONEN 1889: 6, 57, 71; HALLIKAINEN 1998: 15–17). Although besides the general influence of historical and oral traditions, a close reading of these texts also revealed a subtext with connotations to international wilderness narratives. First

Fig. 2. The distribution of archaeological sites in southern Finland in the world's first national atlas, the Atlas of Finland (Suomen kartasto) in 1899 (SUOMEN KARTASTO 1899). Between the most densely populated areas in south-western and south-eastern Finland (Eastern Karelia) is the area of southern Savo, referred as wilderness.

of all it showed links to the literal Georgic tradition,[2] but perhaps most of all, to the Bible. In the following, selected quotations are used in order to portray the influence of such literal ideals on archaeological narratives.

In the Bible and in Judeao-Christian literature the concept of wilderness referred to an area outside the settled area, impervious to holy events. It was considered as a kind of 'non-site', a desert. It was in the wilderness that Jesus met the Tempter, and in the battle between evil and good the wilderness represented wickedness and barbarity (GARRARD 2004: 60–61). In Finnish texts the concept of wilderness as an area hostile to humans, and outside the sphere of the Christian world, can be seen for example in the works of one of Aspelin's contemporaries, Julius Paasonen. According to him:

> "[…] man, rather like an animal than a human being, in dark wilderness and suffering a hard climate, battled for his existence on earth against the beasts and the forces of nature" (PAASONEN 1889: 9; translation mine).

However, while appearing here as a hostile and uncolonized area, the idea of wilderness also carried with it some 'Edenic' underpinnings, as the romantic and nostalgic idea of a past when the first humans lived in harmony with other creatures (CRONON 1996; SLATER 1996). In Aspelin's text, this is expressed in sentences like: "Man, naked and as a creation of God, has begun his life in the great outdoors [...]" (ASPELIN 1885: 1), carrying connotations to Adam and Eve in the Garden of Eden. Respectively, in relation to the Stone Age, he writes that; "the game didn't know to fear humans", so "the hunting was relatively easy" (ASPELIN 1885: 20).

On a deeper level, the Edenic division between the initial state of harmony and the separation of humans from nature in connection to 'the Fall' (SLATER 1996: 115), was a profound element in archaeological texts as it was also used to separate hunter-gatherers from farmers (VILKUNA 1996: 27). Later on, this can be seen more clearly in Tallgren's texts (TALLGREN 1931), in which the relationship towards nature was twofold. During the Stone Age, man was seen as "a man of the forest" – the term carrying connotations to primitivism – while the romantic admiration of wilderness was emphasised in relation to the Iron Age

[2] The 'Georgic trope' has been named after Virgil's (70–29 B.C.) *Georgics*, a poem that describes practical aspects of farming (Greek *georgos,* a farmer) (GARRARD 2004: 109–110).

hunter-farmers who became able to control nature (e.g. TALLGREN 1931: 57, 74, 150–151, 208). Thus, the farming cultures represented the humans after the Fall, to whom God had given domination over the earth: "[…] to till the ground from whence he was taken" (Genesis 3:23). Importantly, through cultivation man came to be seen as integrated with the land, creating an idea of dwelling and historical continuum tied to the 'land of the fore-fathers', and as such it became an object of nationalistic archaeology.

> "It was only recently that the Finns saw the light of history. Those cold northern areas that the Lord gave our ancestors to clear for civilization have demanded a great effort to defeat the rigor of nature" (ASPELIN 1875: foreword; translation mine).

In the quotations above we can also find some links to the literal Georgic tradition deriving from the Old Testament's settler motifs (GARRARD 2004: 108–113). In Georgic tradition the work of man was described as an endless battle against nature, and the hero of the narrative was a hard-working farmer. The influence of this tradition can be seen also in the early interpretations of the Kalevala (1835/1849), the national epic of Finland which was first seen as a heroic battle against nature (LASSILA 2011: 18). In the end of the 19th century, the emphasis shifted to knowledge and intelligence over nature, carrying with it the elements of Enlightment:

> "Man, naked and as a creation of God, begun his life in the great outdoors by sacrificing huge effort in order to study the secrets of nature, the knowledge of which has led him to master nature" (ASPELIN 1885: 1; translation mine).

However, during the next decades the study of the secrets of nature became one of the reasons for the re-evaluation of the concept of wilderness in Finnish archaeology. In the following part, the prehistories of Finland that came to raise the status of the idea of wilderness and give it a prominent position within the national heritage will be discussed more thoroughly.

Wilderness as national heritage – the past in nature

The Aspelinian idea of wilderness as "cold" and "dark" (albeit with Edenic underpinnings), was to be replaced in Finnish archaeology during the coming century. The roots of the opposing trend – the romantic admiration of sublime landscapes – although new in a Finnish context, can be traced back to romantic ideals of the 17th and 18th centuries (*Fig. 3*). From the 19th century onwards, a revitalised admiration of original Finnish nature was integrated with the National Romantic Movement and the building of Finnish identity. This change corresponded with contemporary international trends that raised wilderness as national symbol, especially in the Nordic countries, in Germany and in the United States (HAILA 2003: 179–180; LUKKARINEN 2004; MIKKELI 1992).

In Finland, the building of a national identity based on nature engaged all of society, from authors and artists to scientists and researchers (MIKKELI 1992; PÄIVÄRINNE 2010: 41; SIMOLA 2008b). As a result, nature became part of the national history writing and archaeological narratives. These disciplines were especially inspired by the Kalevala (LÖNNROT 1849; 2010[1888]). Sometimes called "the world's largest forest epic" (SIHVO 2003: 153), the Kalevala came to be interpreted as an expression of a specific Finnish relationship with nature, already in existence during the Prehistoric Era. However, seen in a larger European perspective where forests have had a deep history of meanings from the Antiquity onwards (see HARRISON 1992), the Kalevala is but one out of many epics interpreted in this way.

In the selection of archaeological texts used in this study, *Suomen muinaisuus* (the Antiquity of Finland), written by professor Aarne Michaël Tallgren (1885–1945) in 1931 (TALLGREN 1931), can be seen as a bridge between the changing attitudes towards the idea of wilderness. In his book, the former Christian rhetoric was replaced with a scientific world view that, instead of the Bible, referred to geology in explaining the natural historic past of Finland. This did not mean that the legacies of Christian literature were nowhere to be found. In the form of narratives and motifs, used to give a coherent form and meaning to nature, they persevered. As an example, the Creation in the Book of Genesis was used in order to give coherent form to narratives dealing with the time-span from the Ice Age to the first settlers in Finnish wilderness. It was first used in geology (RAMSAY 1927: 7, 25, 30), and later also in archaeology (e.g. TALLGREN 1931: 18–29) and in biology. As an example of the last, the biologist Kalliola writes:

Fig. 3. The Punkaharju Ridge, Eastern Finland. In Finland, the admiration of sublime landscapes linked with early tourism led to the will to protect them by establishing conservation areas, the first one in 1803 in Punkaharju Ridge when Aleksander I denied forestry and slash-and-burn cultivation on the hills of the ridge. (LEIKOLA 2008: 38) (Photo: I. K. Inha, V99008:63, H. Timgren collection/Metsäkirjasto, Lusto – The Finnish Forest Museum).

> "In the beginning of times, we start with the birth of Finnish bedrock, explore the traces of the great Ice Age, see the organic nature wander to the land uplifting from the ocean, study the life in the old wilderness and finally, follow the conquest of land for cultivation and the evergoing change of nature by culture" (KALLIOLA 1950: 7; translation mine).

Tallgren accepted Aspelin's idea of prehistoric wilderness and deepened it by defining the wilderness geographically. He also dated the phase more precisely to match it with the beginning of agriculture, thereby expressing the basic idea of wilderness as nature distinguished from culture, something set apart from the formation of agricultural landscapes (OELSCHLAEGER 1991: 1–30). The change in the attitude towards wilderness is very clear in the following quotation:

> "The wilderness was colonized step by step and in the solitude of wilderness, the ancient spiritual heritage survived in epic poetry [the Kalevala] creating a bridge between past and present" (TALLGREN 1931: 207; translation mine).

The above mentioned 'spiritual heritage' referred to the national romantic idea of nature as the origin of artistic creativity (PÄIVÄRINNE 2010: 58), a view which was expressed not only in the Kalevala but also in the contemporary Finnish art and in the interpretation of archaeological finds. Tallgren for example (TALLGREN 1931: 192, 206), focused his attention on Eastern Karelia where the majority of the Kalevala's poems were recorded and consequently nominated its Iron Age as the golden age or 'high culture' of Finland, basing his statement on the artistic quality of the finds:

> "[Karelian] stylized art and ornamentation [...] is art at the top of the line: objects are made by the most skilled blacksmiths and craftsmen. However, the wilderness huntsmen were probably much closer to nature and therefore their creations were naturalistic [...] and showed how sensitive they were in describing nature and animals" (TALLGREN 1931: 203, in PÄIVÄRINNE 2010: 58; translation mine).

Besides forming the basis for Finnish creativity and national identity, the idea of wilderness as national heritage also included a more materialistic component.

As is evidenced in the two quotations above, wilderness as a hunting ground for luxurious furs and a place awaiting colonization, draw on the dualistic view of the European wilderness trope, a trope that contradicted romantic admiration with the taming of wilderness for economic wealth (GARRARD 2004: 61–63; VARIS 2003: 312). Archaeology thereby shared the idea of forests as "green gold", a Finnish slogan that referred to the importance of forestry to the developing society (LUKKARINEN 2004: 53–55; PÄIVÄRINNE 2010: 81, 89–90, 116, 119). Similarly, the prehistoric roots of Finland's economic growth as based on its forest were stated in other texts. For example, according to Ella KIVIKOSKI (1961: 277), the forest was "a key to richness" and according to Torsten EDGREN (1993: 270) "the field gave bread but the forest gave richness".

All taken together, the construction of the idea of wilderness in Finnish archaeology was, at the turn of the century, closely related to both international and national conceptions of wilderness and nature. Moreover, the National Romantic ideology united wilderness as well as ancient monuments under the notion of national heritage, both of which needed to be protected by establishing national parks and national museums (PALMGREN 1922: 37–40; see also KRISTIANSEN 1992). This common ideological ground can be read in the 1917 statement, justifying the importance of the Nature Conservation Act by comparing it to the presence of Antiquities Act: "We have legislation for the management of ancient monuments and stray finds. However, we have nothing against the destruction of nature [...]" (AHONEN 1997: chapter 5.1.4.; see PALMGREN 1922: 37–40). Regardless of this common ideological ground, the conservation of natural and cultural heritage diverged into separate administrative branches for decades to come and the synergy between the two was re-established only recently.

Wilderness as otherness

In the two previous parts we followed, by way of expressions and interpretations in early archaeological texts, how the idea of wilderness was modified in different times and contexts; from being seen as a hostile periphery to becoming an admired and idealised landscape, from a national heritage to a resource for economic growth. However, the endurance of this central idea, through sometimes very different situations, brings out the question of the role of the narrator. Therefore, in this part I will briefly discuss the role of the author in relation to the prevailing idea of wilderness as otherness, emanating from the dualistic division between culture and nature (CRONON 1996: 86–89; GARRARD 2004: 69).

An attempt to date the origins of the confrontation between nature and culture has been done by ecolinguist Alwin Fill, who argues that the separation occurred already in the Palaeolithic era and the early stages of the evolution of language, underlining that the human viewpoint to the world constructed nature as "other for me", as opposite to the human "I" (FILL 2010: 177, 183–186). In general, the use of wilderness narratives has been dedicated to farming or modern societies, the latter to which the forested areas represented 'otherness' (CRONON 1996: 87–89; GARRARD 2004: 60–61; HARRISON 1992: 201). Although in the literal wilderness genre, the voice of a farmer can also be read into a pastoral trope influenced by Georgic tradition, as mentioned earlier in regards to Aspelin's texts (GARRARD 2004: 33–58, 108–120). Outside of Finland, the biblical settler motif has also been used for political purposes in other parts of Europe, for example in Norway (OLSEN 2003[1997]: 231) and in Germany (GARRARD 2004: 111–113; VÄYRYNEN 2006: 276–277; see also HUGGAN – TIFFIN 2010; BRUCHAC *et al.* 2010).

So what were the consequences of this 'otherness'? In these literal traditions, the concept of dwelling was most often defined by the peasant way of living, which ideologically denied occupancy and the right to land for hunter-gatherers and nomads (see GARRARD 2004: 108–113). Politically, it also depopulated the wilderness and legitimated the colonialist viewpoint over nature and natives (e.g. BUELL 2005: 66–67; WOBST 2010: 20–22). Respectively, in Finnish archaeology – which was rooted in a tradition of nationalistic history writing – the peasant culture came to symbolize the long continuity of the Finnish nation state, in spite of thousands of years of hunter-gatherer occupation in the wilderness areas (MYLLYNTAUS 2002; see also OLSEN 2003 [1997]: 231).

In the Finnish texts discussed in this paper, the depopulating or 'de-archaeologizing' of the wilderness, despite its indigenous populations, was constructed in two ways. First, the indigenous Saami populations were classified as nomads "wandering in the area", therefore not forming any settlements comparable to those of farming societies. Secondly, the primitivistic view documented already in the early writings of ASPELIN (1885: 30) and PAASONEN (1889: 9), equated the Saami with nature and thus, considered them as archaeologically invisible. This view of indigenous people as underdeveloped extended into archaeological literature, and retained its presence well into the 1970s, or in some cases even later. For example, Lehtosalo-Hilander discussed the material culture of indigenous peoples in the following way:

> "What might these trade-ins have been, as they are so few in the archaeological record? Salt, maybe? Cloth, pretty ribbons? Colourful clothes and ribbons have, though, always been favoured by natives" (LEHTOSALO-HILANDER 1988: 155; translation mine).

Besides being narratives told by modern– or farming communities, the gender of the narrator can also often be identified in these texts, since wilderness narratives almost exclusively concentrate on male activities, thereby effectively marginalising women (GARRARD 2004: 49). Respectively in Finnish archaeology, the activities that were to be found in prehistoric wilderness were those perceived as masculine, such as fur-hunting and "the adventurous sheen of legendary raids" (KIVIKOSKI 1961: 293). In line with this type of reasoning, those areas in which the archaeological record did consist of women's brooches and ornaments, the interpretations nevertheless consigned women to the dwelling and saw them as belonging to the beginnings of colonialism (LEHTOSALO-HILANDER 1984: 329; 1988: 169–170).

> "The women of Haminalahti and Kuusilahti [Iron Age sites] were not temporary acquaintances of wilderness hunters but the housewives of hunting stations [...] During the Viking Age the exploitation of wilderness became systematic so that the wilderness stations, colonized by families, were established in certain crossing points" (LEHTOSALO-HILANDER 1988: 169; translation mine).

To sum up, it is interesting to see that the wilderness narratives cross-sectioned the practice of archaeology also at a more general level, since the political ideas of depopulating– or the ideological notions of a depopulated wilderness, relegated wilderness areas to the periphery of archaeological research for decades. A gradual change in this viewpoint is visible from the 1970s onwards (HUURRE 1979: 124, 143, 172). However, the idea of the wilderness as an area outside the scope of human influence effectively led to the unawareness or disinterest in the cultural heritage within forests and of forests in themselves, contributing to the destruction of archaeological sites by forest machines, a problem that was realised only recently.

Conclusion

Although the environment has been a subject of archaeological research in Finland from the 1920s, it was the overall politicization of nature that occurred in the 1970s that formed environmental issues into a coherent whole (see TRIGGER 2008: 258). This change can be seen in the increasing amount of environmentally oriented publications, 90% of which were published the 1970s onwards (KIRKINEN 2007: 69). However, the cultural constructions of nature, and the underlying human-nature relationship – circulated and recycled in the form of narratives – can be found in Finnish archaeological publications from the very beginnings of the discipline.

Based on the discussions in this paper, it can be said that the early adoption of wilderness narratives into Finnish archaeology was most notably inspired by Judaeo-Christian texts, updated in intervals by contemporary literal trends and by historically known traditions of wilderness-use. Later on, the changes in society brought forward new conceptions of (and attitudes towards) nature, such as the ones promoted by nationalism, national romanticism, and environmentalism. These, in turn, influenced the wilderness genre and archaeological nature writing as well. While reading the texts in chronological order it was also easy to detect how they were engaged in dialogue with previous works. Along with the accumulation of the archaeological record over several decades, the idea of wilderness was constructed in a more detailed form. For example in the 1930s, Tallgren localized the mythical Finnish forests geographically and dated the exact beginning of wilderness culture to the Bronze Age, so as to coincide with the beginning of agriculture.

Most importantly, as a genre adapted to a Finnish context, wilderness narratives formed a long-lasting hypothesis against which the interpretations of archaeologically low-intensive areas were constructed. In Finland, the different features and outcomes in archaeological interpretations can be listed as follows:

1) Wilderness as an uncolonized area. This statement had a political agenda in the sense that it legitimated the farmer colonization and land-use in "wild" or forested areas, regardless of indigenous populations. This ideology finds historical parallels in situations such as when Finnish colonists in the north didn't recognise and respect the Saami right to own land (KORPIJAAKKO 1989). This interpretation or use of wilderness is also bound to nationalistic archaeology, which raised the Finnish peasant

culture and created it as a symbol of the continuity of the nation. In doing so, it marginalised the archaeology of wilderness areas.

2) Wilderness as pristine. The idea of wilderness as pure nature corresponded well with the myth about hunter-gatherers living in harmony with nature. This, in turn, marginalised and made abundant the study of early human impact on nature. However, a recent change in paradigms has led to the uncovering of the previously unknown, long-lasting land-use legacies of early human activity as well as the human impact on the formation of biologically valued biotopes (VAINIO *et al.* 2001).
3) Wilderness as wild. The idea of wilderness as an area outside of civilization separated the 'domesticated species' from those of the forests. As an example, in Eastern Finnish wilderness areas, the activity of cattle breeding was long considered unlikely in spite of the osteological record showing the existence of domesticated species already in the 1930s. Similarly, the palaeoecological evidence, showing sporadic cultivation in wilderness areas, was interpreted almost entirely as a result of colonization processes.
4) Wilderness as a stage for male culture. In archaeological texts, the wilderness was viewed almost entirely from the males' point of view, as a goal for their travels and hunting trips, journeys carrying an aura of danger and adventure. In this tradition women were outsiders as they represented the dwelling. As a result, any type of female wilderness culture was excluded in archaeological wilderness writing, with only a few short exceptions.

The impacts of the tropes and traditions connected to wilderness narratives on archaeological texts are evident, but archaeological constructions of the past have also affected the idea of the wilderness. First, the 'blanks' or white areas on archaeological distribution maps were used to prove the continuous vacancy of such lands, and thereby also the originality or pristine character of nature in the wilderness. Secondly, in the attempts to preserve or museumize the remnants of the past for future generations, the natural heritage became paralleled with archaeological remains, which in turn motivated the development of nature protection in Finland (PALMGREN 1922). As a result of all of these developments, the study of wilderness – nowadays referred as woodland archaeology – needs to incorporate alternative views that enable new ways of knowing the past. Important steps towards this have already been taken, as the FINNISH FOREST AND PARK

SERVICE (2012) is conducting large surveys in state-owned forests and mapping cultural heritage, from the Stone Age to the mid of the 20th century. The National Board of Antiquities has, in turn, promoted the protection of archaeological sites in private-owned forests, in order to minimize the threats caused by forestry (VÄSTERBOTTENS MUSEUM 2009). In the future, the decolonizing of the wilderness – i.e. the disengagement with the biased colonial view on wilderness areas as representing otherness, vacancy or unaffected by humans – can help us to free woodland archaeology from the baggage of history (see e.g. WOBST 2010: 22–23; BRUCHAC 2010: 367).

References

AHONEN, SANNA (1997), *Mitä on suojeltu, kun on suojeltu luontoa? Käsitehistoriallinen tarkastelu suomalaisesta luonnonsuojelusta välillä 1880–1983*. Master thesis. Univ. of Helsinki, Faculty of Agriculture and Forestry. (http://ethesis.helsinki.fi/julkaisut/maa/limno/pg/ahonen/ [last updated 2001]).

ASPELIN, JOHANNES REINHOLD (1875), *Suomalais-ugrilaisen muinaistutkinnon alkeita*. Helsinki.

ASPELIN, JOHANNES REINHOLD (1880), *Muinaisjäännöksiä Suomen suvun asumus-aloilta. Rauta-aika*. Helsinki.

ASPELIN, JOHANNES REINHOLD (1885), *Suomen asukkaat pakanuuden aikana*. Helsinki.

BRIGGS, JOHN – KATHERINE SPIELMANN – HOSKI SCHAAFSMA – KEITH KINTIGH – MELISSA KRUSE – KARI MOREHOUSE – KAREN SCHOLLMEYER (2006), Why ecology needs archaeologists and archaeology needs ecologists. *Frontiers in Ecology and the Environment* 4/4, 180–188.

BRUCHAC, MARGARET (2010), Imagining the future of indigenous archaeologies. In: Margaret Bruchac – Siobhan Hart – Martin Wobst 2010, 363–368.

BRUCHAC, MARGARET – SIOBHAN HART – MARTIN WOBST (eds) (2010), *Indigenous archaeologies. Reader on decolonization.* Walnut Creek.

BUELL, LAWRENCE (1995), *The environmental imagination. Thoreau, nature writing, and the formation of American culture.* Cambridge – London.

BUELL, LAWRENCE (2005), *The future of environmental criticism: Environmental crisis and literary imagination*. Malden.

CRONON, WILLIAM (1996), The trouble with wilderness; or, getting back to the wrong nature. In: William Cronon (ed.), *Uncommon ground. Rethinking the human place in nature*. New York – London. 91–113.

EDGREN, TORSTEN (1993), Den förhistoriska tiden. In: Märtha Norrback (ed.), *Finlands historia* 1. Ekenäs. 11–270.

FILL, ALWIN (2010), *The language impact. Evolution — system — discourse*. London.

FINNISH ENVIRONMENT INSTITUTE (2013), More accurate information available about urban and rural areas. (http://www.syke.fi/en-US/SYKE_Info/Communications_material/Press_releases/More_accurate_information_available_abou(14636) [17.06.2013]).

FINNISH FOREST AND PARK SERVICE (2012), Field survey of cultural heritage sites in state-owned forests of Finland. (http://www.metsa.fi/sivustot/metsa/en/Projects/CulturalHeritageFieldSurvey/Sivut/default.aspx [Last updated 30.10.2012]).

FINNISH FOREST ASSOCIATION (2012), Finland is the most forested country in Europe. (http://www.forest.fi/smyforest/foresteng.nsf/allbyid/BE3C5576C911F822C2256F3100418AFD?Opendocument [01.06.2012]).

FOSTER, DAVID – FREDERICK SWANSON – JOHN ABER – INGRID BURKE – NICHOLAS BROKAW – DAVID TILMAN – ALAN KNAPP (2003), The importance of land-use legacies to ecology and conservation. *BioScience* 53.1, 77–88.

GARRARD, GREG (2004), *Ecocriticism*. London.

GLOTFELTY, CHERYLL (1996), Introduction. In: Cheryll Glotfelty – Harold Fromm (eds), *The ecocriticism reader: landmarks in literary ecology*. Athens. xv–xxxvii.

HAILA, YRJÖ (2003), 'Erämaa' ja ympäristöajattelun monimuotoisuus. In: Yrjö Haila – Ville Lähde – Mikko Lehtonen (eds), *Luonnon politiikka*. Tampere. 174–204.

HALLIKAINEN, VILLE (1998), *The Finnish wilderness experience*. Metsäntutkimuslaitoksen tiedonantoja 711. Rovaniemi.

HARRISON, ROBERT (1992), *Forests. The shadow of civilization*. Chicago – London.

HAYASHIDA, FRANCES (2005), Archaeology, ecological history, and conservation. *Annual Review of Anthropology* 34, 43–65.

HEISE, URSULA (1999), Science and ecocriticism. *Mots Pluriels* 11/Septembre 1999. (http://motspluriels.arts.uwa.edu.au/MP1199ukh.html [27.07.2013]).

HOWARTH, WILLIAM (1996), Some principles of ecocriticism. In: Cheryll Glotfelty – Harold Fromm (eds), *The ecocriticism reader: Landmarks in literary ecology*. Athens. 69–91.

HUGGAN, GRAHAM – HELEN TIFFIN (2010), *Poscolonial ecocriticism. Literature, animals, environment*. London – New York.

HUURRE, MATTI (1979), *9000 vuotta Suomen esihistoriaa*. Helsinki.

JOYCE, ROSEMARY – CAROLYN GUYER – MICHAEL JOYCE (2002), Voices carry outside the discipline. In: Rosemary Joyce – Robert W. Preucel – Jeanne Lopiparo – Carolyn Guyer – Michael Joyce (eds), *The languages of archaeology. Dialogue, narrative, and writing*. Cornwall. 100–132.

KALLIOLA, REINO (1950), *Suomen kaunis luonto*. Porvoo.

KIRKINEN, TUIJA (2007), Esihistoria ympäristöhuolen aikakaudella — arkeologian ja luonnonsuojelun väliset yhteydet. In: Teemu Mökkönen – Sirkka-Liisa Seppälä (eds), *Arkeologian yhteiskunnallinen merkitys & Arkeologisia ilonaiheita*. Arkeologipäivät 2007. Helsinki. 64–73.

KIVIKOSKI, ELLA (1961), *Suomen esihistoria*. Suomen historia 1. Porvoo.

KIVIKOSKI, ELLA (1964), *Finlands förhistoria*. Stockholm.

KIVIKOSKI, ELLA (1967), *Finland*. London.

KORPIJAAKKO, KAISA (1989), *Saamelaisten oikeusasemasta Ruotsi-Suomessa: Oikeushistoriallinen tutkimus Länsi-Pohjan Lapin maankäyttöoloista ja-oikeuksista ennen 1700-luvun puoliväliä*. Doctoral thesis. Univ. of Lapland. Helsinki.

KRISTIANSEN, KRISTIAN (1992), From romanticism, through antiquarianism, to an historical view of nature: the case of Denmark. In: Lesley Macinnes – Caroline R. Wickham-Jones (eds), *All natural things: archaeology and the green debate*. Oxbow monograph 21. Oxford. 52–64.

LASSILA, PERTTI (2011), *Metsän autuus. Luonto suomalaisessa kirjallisuudessa 1700–1950*. Suomalaisen Kirjallisuuden Seuran Toimituksia 1344. Helsinki.

LEHTOSALO-HILANDER, PIRKKO-LIISA (1984), Keski- ja myöhäisrautakausi. In: Yrjö Blomstedt – Veikko Anttila – Jukka Tarkka (eds), *Suomen historia 1*. Espoo. 250–405.

LEHTOSALO-HILANDER, PIRKKO-LIISA (1988), Esihistorian vuosituhannet Savon alueella. In: Martti Ruuth – Kauko Pirinen – Pirkko-Liisa Lehtosalo-Hilander (eds), *Savon historia I*. Kuopio. 12–264.

LEIKOLA, ANTO (2008), Luonnonsuojelun tulo Suomeen. In: Helena Telkänranta (ed.), *Laulujoutsenen perintö. Suomalaisen ympäristöliikkeen taival*. Porvoo. 34–41.

LOVE, GLEN (1999), Ecocriticism and science: Towards consilience? *New Literary History* 30.3, 561–576.

LOVE, GLEN (2003), *Practical ecocriticism. Literature, biology, and the environment*. Charlottesville – London.

LÖNNROT, ELIAS (ed.) (1849), *Kalevala*. Suomalaisen Kirjallisuuden Seuran Toimituksia 14. Helsinki.

LÖNNROT, ELIAS (2010 [1888]), *The Kalevala*. English translation by John Martin Crawford (http://www.gutenberg.org/cache/epub/5186/pg5186.html [27.07.2013]).

LUKKARINEN, VILLE (2004), Kansallisen maiseman vertauskuvallisuus ja ympäristön tila. In: Ville Lukkarinen – Annika Waenerberg (eds), *Suomi-kuvasta mielenmaisemaan. Kansallismaisemat 1800–1900-luvun vaihteen maalaustaiteessa*. Suomalaisen Kirjallisuuden Seuran Toimituksia 965. Helsinki. 20–91.

MIKKELI, HEIKKI (1992), Metsäturkki ja sen jurot parturit. Näkemyksiä metsäluonnon ja kansanluonteen suhteesta 1800–1900-luvulla. *Historiallinen Aikakauskirja* 92.3, 200–214.

MYLLYNTAUS, TIMO (2002), Writing about the past with green ink: The emergence of Finnish environmental history. In: *H-Net Humanities Online, Historiography series in global environmental history* (http://www.h-net.org/~environ/historiography/finland.htm [24.07.2013]). Also in: Erland Mårald – Christer Nordlund (eds) (2003), *Värna, vårda, värdera: miljöhistoriska aspekter och aspekter på miljöhistoria*, Landskapet som arena 5. Umeå. 177–200.

NORDMAN, CARL AXEL (1924), *Karelska järnåldersstudier*. Suomen Muinaismuistoyhdistyksen aikakauskirja 34.3. Helsinki.

OELSCHLAEGER, MAX (1991), *The idea of wilderness*. New Haven – London.

OLSEN, BJØRNAR (2003 [1997]), *Från ting till text. Teoretiska prespektiv i arkeologisk forskning*. Lund.

PAASONEN, JULIUS (1889), *Muinaisjäännöksiä ja muistoja Mikkelin kihlakunnassa*. Luetteloja Suomen muinaisjäännöksistä 15. Helsinki.

PALMGREN, ROLF (1922), *Luonnonsuojelu ja kulttuuri 1–2*. Helsinki.

PÄIVÄRINNE, TIINA (2010), *Luonto, tiede ja teknologia. Kansanvalistuksen Suomi-kuva 1870–1920*. Bidrag till kännedom av Finlands natur och folk 183. Helsiki.

RAMSAY, WILHELM (1927), Silmäys Suomen syntyyn. In: Waldemar Ruin, *Kutsu niihin maisterin- ja tohtorinpromotsioneihin jotka Helsingin yliopiston filosofinen tiedekunta tavanmukaisin juhlallisuuksin viettää yliopiston juhlasalissa toukokuun 31 p. 1927*. k: lo 11 a.p. [s.n.], Helsinki. 5–30.

SALO, UNTO (1984), Pronssikausi ja rautakauden alku. In: Yrjö Blomstedt – Veikko Anttila – Jukka Tarkka (eds), *Suomen historia 1*. Espoo. 98–249.

SIHVO, HANNES (2003), Vihreän kullan maa. In: Yrjö Sepänmaa – Liisa Heikkilä-Palo – Virpi Kaukio (eds), *Metsään mieleni*. Hämeenlinna. 151–161.

SIIRIÄINEN, ARI (1987), Suomen esihistoria. In: Seppo Zetterberg (ed.), *Suomen historian pikkujättiläinen*. Porvoo. 11–39.

SIMMONS, IAN (1993), *Interpreting nature. Cultural constructions of the environment*. London – New York.

SIMOLA, HEIKKI (2008a), Miten Suomen luonto löydettiin. In: Helena Telkänranta (ed.), *Laulujoutsenen perintö. Suomalaisen ympäristöliikkeen taival*. Porvoo. 18–21.

SIMOLA, HEIKKI (2008b), Taiteilijoiden luonto ja kansallisuusaate. In: Helena Telkänranta (ed.), *Laulujoutsenen perintö. Suomalaisen ympäristöliikkeen taival*. Porvoo. 24–31.

SLATER, CANDACE (1996), Amazonia as Edenic narrative. In: William Cronon (ed.), *Uncommon ground. Rethinking the human place in nature*. New York – London. 114–131.

SUOMEN KARTASTO (1899). Suomen maantieteellinen seura. Helsinki.

TAAVITSAINEN, JUSSI-PEKKA – HEIKKI SIMOLA – ELISABETH GRÖNLUND (1998), Cultivation history beyond the periphery: Early agriculture in the North European boreal forest. *Journal of World Prehistory* 12.2, 199–253.

TALLGREN, AARNE (1931), *Suomen muinaisuus*. Suomen historia I. Porvoo.

TRIGGER, BRUCE (2008), *A history of archaeological thought*. (2nd edition). Cambridge.

VAINIO, MAARIT – HANNELE KEKÄLÄINEN – AULIKKI ALANEN – JUHA PYKÄLÄ (2001), *Suomen perinnebiotoopit: Perinnemaisemaprojektin valtakunnallinen loppuraportti*. Suomen ympäristö 527. Helsinki.

VARIS, MARKKU (2003), *Ikävä erätön ilta. Suomalainen eräkirjallisuus*. Suomalaisen Kirjallisuuden Seuran Toimituksia 939. Helsinki.

VILKUNA, JANNE (1996), Siperiasta vai? Uusia tulkintoja suomalaisten alkuperästä. In: Heikki Roiko-Jokela (ed.), *Siperiasta siirtoväkeen: murrosaikoja ja käännekohtia Suomen historiassa*. Jyväskylä. 7 30.

VÄSTERBOTTENS MUSEUM (2009), *SKAIK – Skogens KulturArv I Kvarkenregionen*. (http://www.skaik.eu/ [07.01.2013]).

VÄYRYNEN, KARI (2006), *Ympäristöfilosofian historia maaäitimyytistä Marxiin*. Tampere.

WOBST, MARTIN (2010), Indigenous archaeologies: a worldwide perspective on human materialities and human rights. In: Margaret Bruchac – Siobhan Hart – Martin Wobst (eds), *Indigenous archaeologies. Reader on decolonization*. Walnut Creek. 17–27.

A report from a person who appropriates archaeologists' narratives for the public

MICHAEL A. CREMO

Abstract:
One view of communication between archaeologists and the public is that archaeologists have expert knowledge which for their own purposes they communicate in suitable form to the public, making use of mediators, such as reporters. This view expresses only part of the complex relationship between the public and archaeologists. Mediators and the public are not simply passive receivers of knowledge emanating from archeologists. Some persons are actively appropriating archaeological narratives and information, which they transform and communicate for their own purposes to the public. I am one such person. In my books, I appropriate archaeological narratives and information, configure them to support my own views about human origins and antiquity, and communicate the product to the public, and back to the archaeologists (in papers presented at meetings of the World Archaeological Congress, the European Association of Archaeologists, and the International Union for Prehistoric and Protohistoric Sciences). In this paper, I explore the motives and methods involved in my public appropriation of archaeological narratives and information, in the context of my relationship with the community of archaeologists.

I am not a professional archaeologist, but I am an associate member of the European Association of Archaeologists (EAA). Early in 2012, I received the call for papers for the 2012 meeting of the EAA in Helsinki, Finland. So I went to the conference website and searched through the list of accepted sessions, to see if I could find one that related to my particular interests. None of them attracted me, until I came upon this session: "Public Appropriations of Archaeologists Narratives", organized by Thomas Meier and Elisabeth Niklasson. As soon as I read the session title, I experienced a shock of self-recognition. "Yes," I said to myself, "I am appropriating archaeologists' narratives." So I submitted the abstract you see above. It was accepted by the academic committee of the conference, so I wrote a paper. I intended it to be not simply a paper that happens to fit within the boundaries of the session, but rather a paper that engages directly and specifically with the session organizers' abstract for the whole session (MEIER – NIKLASSON 2012). In Helsinki, I gave a powerpoint presentation in which I briefly summarized some of the main points I made in the paper I had written. Based on the questions and comments I received on my presentation, and taking into account the other presentations in the session, and the exchanges related to them, I have prepared this present paper, which I prefer to call a report.

Archaeologists, mediators, and publics

In the model suggested by Meier and Niklasson, there are three elements:

(1) archaeologists,
(2) mediators, such as reporters for print and electronic media, and
(3) the public, or publics.

According to one common view, archaeologists, using suitable mediators, transmit messages to members of the public, who, ideally, understand and accept the messages just as the archaeologists intend them to be understood and accepted. Meier and Niklasson note, however, that archaeologists are sometimes disappointed to see that, in the process of mediated transmission and public reception, their messages are appropriated and transformed in (to them) undesirable ways. Adopting a critical stance toward this attitude, they state that: "The paternalistic guidance of the public towards the academically sanctioned truth, as endorsed by modernity, has meant that these appropriations have consistently been disregarded and deemed useless." Instead, they advocate a different attitude, which involves regarding at least some of these appropriations as "productive" (Meier – Niklasson 2012: 73).

Positioning myself as a mediator

Where do I fit into the categories outlined by Meier and Niklasson? I believe the three categories (archaeologists, mediators, and the public) are not mutually exclusive. Meier – Niklasson (2012: 73) themselves hinted at this in their session abstract, where they said: "When talking about 'the public' we [archaeologists] often routinely construct 'them' as a distinguishable group separate from 'us'." Meier and Niklasson imply they are not always distinguishable groups, and I accept that implicit conclusion. The same person may in different situations be archaeologist, mediator, or member of a public, or, to use other terms, a generator of archaeological narratives, transmitter of archaeological narratives, or receiver of archaeological narratives. Despite the fuzzy, overlapping nature of the categories, I still find them useful. So in terms of those categories, how do I position myself for the purposes of this report? For the moment, I agree to characterize myself as a mediator.

However, I am not a passive mediator chosen by archaeologists for the purpose of reliably transmitting their narratives to the public. I am not, for example, a science reporter for a mainstream media outlet who is on the list to receive press releases from professional archaeologists. Instead, I am, for the purpose of this part of my report, an independently active mediator. I willfully appropriate archaeologists' narratives and information that I find useful for my purposes, which are not always the same as those of the archaeologists who were the originators of the narratives and information.

My transcultural identity

If one wishes to understand how and why I appropriate and transform archaeologists' narratives in my function as a mediator who communicates with the public, one has to take into account my transcultural identity. I use the term "transcultural" as it is defined and explained by the German philosopher Wolfgang WELSCH (1999). He prefers the term transcultural to multicultural, which implies distinct cultures interacting within a polity. Welsch says (WELSCH 1999: 197):

> "Cultures de facto no longer have the insinuated form of homogeneity and separateness. They have instead assumed a new form, which is to be called *transcultural* insofar that it *passes through* classical cultural boundaries. Cultural conditions today are largely characterized by mixes and permeations."

Transculturality contributes to individual identity formation. Welsch says (WELSCH 1999: 198):

> "Transculturality is gaining ground moreover not only on the macro-cultural level, but also on the individual's micro-level. For most of us, multiple cultural connections are decisive in terms of our cultural formation. We are cultural hybrids."

So I am, as are many of you, a transcultural person, a cultural hybrid. I was born in the United States. My father was a military intelligence officer. I lived in different countries, growing up among people involved in diplomacy, government, and the military. In those days, I was part of American Cold War culture. In the fall of 1966, I entered the school of international affairs at the George Washington University in Washington, D.C. But my career ambitions and commitment to modernist university education did not survive my exposure to the "counterculture" and the movement against the Vietnam war. I pursued another path. I was especially attracted to the spiritual culture of India, and eventually joined the International Society for Krishna Consciousness. I became an initiated disciple of the Indian guru who founded the Society, and I practiced a system of yoga called *bhakti* yoga, the yoga of devotion.

A key part of *bhakti* yoga is using one's talents and abilities for spiritual purposes. I had talents for research and writing, so I engaged them as part of my yoga. I channeled my research impulses into studying the texts of ancient India, collectively known as the Vedic literature. Among the Vedic texts are the *Puranas*, the historical texts. In *Puranas*, especially the *Bhagavata Purana*, I found accounts of extreme human antiquity, accounts of human populations that existed millions of years ago, going back to the very beginnings of life on earth. That contradicted the accounts of human origins I had learned from my teachers in secondary school and university. Those conventional accounts placed the origin of the modern human species less than 200,000 years ago.

So how did I deal with that contradiction? I chose to hypothesize that the Puranic accounts of extreme human antiquity, if accepted as true, might be consistent with some archaeological evidence. That choice led me into a deep study of the history of archaeology, to see if there was any evidence at all for extreme human antiquity.

My appropriations of archaeologists' narratives

My first step in my study of the history of archaeology was to carefully examine the current secondary literature (textbooks, survey studies, etc.). There I found no explicit evidence for extreme human antiquity. I then decided to look beyond the current secondary literature. I began to study the primary scientific literature (reports by original researchers, published in the professional scientific literature), both past and present. Because I had a reading knowledge of many of the major European languages, I did not confine myself to reports in English. In this search, which lasted eight years, I encountered many reports of archaeological evidence for extreme human antiquity, and I appropriated them for my purposes (CREMO – THOMPSON 1993).

What exactly did I appropriate? Among other things, I appropriated the archaeological narratives of Tertiary Man, from the late nineteenth century and early twentieth century (e.g. RIBEIRO 1873; WHITNEY 1880; RUTOT 1907), and deployed them to problematize the foundational narrative of modern prehistoric archaeology – the Late Middle Pleistocene appearance of the modern human species. In the case of the narratives of Tertiary Man, I did not have to engage in much transformation of the narratives, because many of the originators of these narratives accepted extreme human antiquity.

I have also appropriated narratives from more recent archaeologists (and other scientists, such as geologists and physical anthropologists, involved in archaeological discoveries). In most cases, I have had to reconfigure the narratives and information generated by modern archaeologists to incorporate them into my project. For example, in 1979, Mary Leakey announced the discovery of footprints at Laetoli, in Tanzania. The footprints were found in solidified volcanic ash 3.6 million years old. In her original report, Mary LEAKEY (1979: 453) said the form of the foot that made the prints "was exactly the same as ours." She did not, of course, believe humans like us made the footprints. She proposed they were made by some type of human precursor, some Pliocene hominin, who had feet just like modern human feet. I appropriated Leakey's narrative, but reconfigured it to support my own narrative of extreme human antiquity. I proposed that one could not exclude the possibility that the footprints were made by humans like us, rather than a precursor: "But why not a creature with fully modern feet and fully modern body? There is nothing in the footprints that rules this out" (CREMO – THOMPSON 1993: 743).

It is not my purpose in this report to defend the validity of my appropriations against various objections and criticisms. I am just reporting what I have done. I collected hundreds of cases of archaeological material, from all periods of archaeological research, in my book *Forbidden Archeology* (CREMO – THOMPSON 1993; I deliberately used the American spelling of archaeology). So, briefly, that is how, as a mediator, I independently appropriated archaeologists' narratives, sometimes reconfiguring them, to support my own favored narrative of extreme human antiquity.

One might ask about the nature of my commitment to my own favored narrative. Keep in mind that I am entering the arena of contemporary archaeology as a member of a different knowledge tradition. According to this knowledge tradition, there are three main sources of knowledge: *pratyaksha* (sense evidence), *anumana* (logical inference), and *shabda* (revelation, recorded in texts, in this case the *Puranas*). Among the three, revelation is considered superior and sufficient. Contemporary archaeology, for the most part, does not acknowledge scriptural revelation as an independent source of real knowledge about human origins and antiquity. It recognizes primarily sense evidence and whatever can be logically inferred from such sense evidence. Among members of my own knowledge tradition, I could answer questions about human origins and antiquity simply by giving statements from the *Puranas*. I cannot do that when I come as a guest among contemporary archaeologists. What I can do, in terms of the practices of their knowledge tradition, is make proposals having their source in the *Puranas,* offering in support of those proposals whatever archaeological evidence and reasoning I can.

After I presented my powerpoint talk at the EAA meeting in Helsinki, one of the session organizers, Elisabeth Niklasson, asked me if I would ever be prepared to change my Puranic conclusions (such as extreme human antiquity), based on archaeological evidence that came to my attention. I thought about that. What, for example, would I do if I were compelled to admit that all of the cases of archaeological evidence for extreme human antiquity that I refer to in my work are invalid, disproved. My answer was that I might not change my conclusions, because they are based on what I regard as a higher source of knowledge (the revelation of the *Puranas*), but I might stop trying to justify these conclusions outside the circles of those who share my acceptance of this source of revelation. However, as I will argue later in this report, even if I am not ready to change fundamental aspects of my account of the past, my intervening in the field may still have some positive effects on archaeology as a discipline.

One of my publics: professional archaeologists

Now let us consider my audiences, my publics. One of my publics is in fact the community of professional archaeologists. Why? While I do not think that professional archaeologists have a monopoly on authority in questions relating to human origins and antiquity, because of their focused, disciplined study of the physical remains of the past they do have some considerable degree of authority in influential circles, and are part of the whole constellation of groups and individuals concerned with such things. And I want to address the whole constellation, so for that reason I include professional archaeologists among my publics.

I appropriate archaeologists' narratives, past and present, and reflect them back to the living community of archaeologists (and other scientists concerned with human origins), for the purpose of problematizing the foundational evolutionary narrative of prehistoric archaeology, or at least the exclusivity of its position in archaeology. An even deeper purpose is to contribute to configuring archaeology transculturally, making more intellectual space for my Puranic archaeological hybrid. In my work, I have been very open about my transcultural identity and the transcultural inspiration for my project. It may seem somewhat paradoxical that I come into archaeology with some non-negotiable truth claims while at the same time focusing, in postmodernist fashion, on archaeology as a changeable social construction, which appears to entail a relativist conception of truth. I do not see relativism as the ultimate position. Rather, I see the postmodern trend in archaeology as opening the discussion on what is ultimately true to a plurality of approaches, out of which some consensus might ultimately emerge. And if it does not, then the openness to a plurality of approaches remains appropriate in archaeology.

How did I initially establish communication with the community of archaeologists? First of all I configured my book *Forbidden Archeology* to resemble an academic archaeological text. Then when it was published, I sent it out for review in professional journals of archaeology, anthropology, and history of science. How was the book received? I will divide my consideration of this question into two parts: (1) how the bare appropriation by me of archaeological narratives and information was received, and (2) how the transcultural element of my project was received.

My appropriation of archaeologists' narratives and information inspired various reactions, some more negative and some more positive. Many reviews were published, but I will mention only a couple of representative examples. On the negative side, anthropologist Jonathan MARKS (1994: 141) dismissively called the book "a veritable cornucopia of dreck." On the more positive side, archaeologist Tim MURRAY (1995: 379) said:

> "I have no doubt that there will be some who will read this book and profit from it. Certainly it provides the historian of archaeology with a useful compendium of case studies in the history and sociology of scientific knowledge, which can be used to foster debate within archaeology about how to describe the epistemology of one's discipline."

Turning to the transcultural aspect, again there were negative and positive reactions. Jonathan MARKS (1994: 141) characterized the book as "Hinduoid creationist drivel." Others were less critical of the book's Vedic inspiration. MURRAY (1995: 379) said:

> "The 'dominant paradigm' [in archaeology] has changed and is changing, and practitioners openly debate issues which go right to the conceptual core of the discipline. Whether the Vedas have a role to play in this is up to the individual scientists concerned."

I regard that as an enlightened opinion. I am not suggesting that Murray endorses my conclusions. He does not, but he does see some value in my work for archaeologists and historians of archaeology.[1]

Another way I communicate with the community of archaeologists is by presenting papers, which contain my appropriations of archaeologists' narratives, at archaeological conferences. The first time I did this was in 1994, when I presented a paper titled "Puranic Time and the Archaeological Record" in a session on time and archaeology at the meeting of the World Archaeological Congress held in New Delhi. That paper later appeared in a peer-reviewed conference proceedings volume published by Routledge (CREMO 1999). Over the years, I have presented many other papers at various WAC congresses and meetings of other archaeological organizations, including the European Association of Archaeologists and the European Association for South Asian Archaeology and Art. I have also presented papers on my transcultural archaeological work at meetings of historians of science. Furthermore, I have given invited papers about my work at scientific institutes, including archaeological institutes.

The complete texts of many of these papers can be found in a recently published book of mine (CREMO 2012a). Versions of several of the papers have

---

[1] For more reactions to my work from professional archeologists and other academics, see my book *Forbidden Archeology's Impact* (CREMO 1998). About this book, a scholar of public understanding of science (LOCKE 1999) wrote: "Cremo has provided here a resource of considerable richness and value to analysts of public understanding [of science]. [...] It should also make a useful teaching resource as one of the best-documented case studies of 'science wars', and raising a wide range of issues covering aspects of 'knowledge transfer' in a manner sure to be provocative in the classroom."

been published in the professional literature (e.g. CREMO 1999; CREMO 2002; CREMO 2008; CREMO 2009).[2]

In addition to presenting papers at major international conference and scientific institutions, I also regularly lecture at universities throughout the world. The very fact that I am able to give such talks I take as a positive development, given that my practice of appropriating archaeologists' narratives in support of Puranic accounts of extreme human antiquity is well known. Reactions to my public lectures at conferences, universities, and scientific institutions are similar to published reactions to my books – some negative reactions, some acknowledgments of positive aspects of my work. I will give an example of each. In 2003, I spoke at the anthropology department of the Russian Academy of Sciences in Moscow. One of the anthropologists present said during the question session after my talk, "I have not read your book, but I am sure that everything in it must be a mistake or a hoax" (CREMO 2010: 62). In 2002, I spoke at the Institute of Archaeology of the Ukrainian Academy of Sciences in Kiev. In an article I wrote about the talk, I recalled (CREMO 2010: 52):

> "One of the archeologists accompanied me out to my car. He said he very much appreciated my talk. He said that some years ago, the archeologists at the institute would have weekly sessions in which they deeply questioned the fundamental assumptions underlying their practice of archeology, but over the years these discussions had stopped. He said that my work made a contribution to getting archeologists thinking about these things again."

Another way that I connect with archaeologists is through my research activities, which take me into archaeological collections in the storerooms of museums, into archival collections of documents related to archaeological discoveries, and into the field to visit archaeological sites. My experiences with archaeologists in these settings have been positive. For example, when I visited the Royal Museum of Natural Sciences in Brussels, Belgium, to study the eolith collection of A. Rutot from Boncelles (regarded by him as Tertiary), archaeologists

---

[2] In this report, I am not going to include a discussion of my presentation of a consciousness-based Vedic alternative to the conventional theories of human origins. A developed presentation of that alternative can be found in my book *Human Devolution* (CREMO 2004). In this paper, I confine myself to a discussion of my work on archaeological evidence for extreme human antiquity.

were polite and helpful, although they did express some mild puzzlement about why I was interested in a collection they regarded as "false".

Positioning myself as a kind of archaeologist

Given my presence in institutions, sites, and publications connected with professional archaeology, let me suggest another way I can locate myself in the categories of Meier and Niklasson (archaeologist, mediator, public). Provisionally, I first located myself as a mediator – not as a member of the public, and not as an archaeologist. This latter negative identification (not an archaeologist) at first glance seems reasonable. I do not have a university degree in archaeology. I am not employed as a professional archaeologist. However, there are theoretical developments in the world of archaeology that might allow me to identify myself as a kind of archaeologist, and not only as a mediator. Some archaeologists propose that there is not just one archaeology; rather there are many archaeologies. Reflecting this understanding, the journal of the World Archaeological Congress is called *Archaeologies*. The WAC is famously open to indigenous archaeological perspectives, which include, for example, traditional Australian aboriginal and Native American Indian perspectives. One might call this an attempt at multiculturality in archaeology. But what about modern transcultural hybrid perspectives such as mine? In this regard, archaeologist Cornelius HOLTORF (2006: 90–91) wrote about WAC:

> "If non-Western [indigenous] groups are made welcome to represent their accounts of the past or archaeological sites according to their own epistemologies and intellectual traditions, it is difficult to see how the same courtesy could be denied to traditions, both within and outside the Western world, that may be similarly different from what is expected within the academic discipline of archaeology [...]. There cannot be different worlds of Western and non-Western or hegemonic and suppressed archaeologies, and there cannot likewise be different worlds of academic and amateur or orthodox and alternative archaeologies."

In 2007, I gave a paper in a session on alternative archaeologies at the EAA annual meeting in Zadar, Croatia. That paper is included in a recent volume titled *From Archaeology to Archaeologies: The "Other" Past*, edited by the session organizers (SIMANDIRAKI-GRIMSHAW – STEFANOU 2012). So this current Helsinki paper is not the first one in which I have reflexively addressed, at the EAA, my positioning in the world of archaeology, or, perhaps better said, in the world of archaeologies. In my Zadar paper (CREMO 2012b), I positioned myself as an "alternative archaeologist".

I would say this is reasonable, because one of the elements in my transcultural hybridization is in fact a substantial element of what we might call "traditional" academic archaeology. I can give a convincing imitation of an academic archaeological paper, convincing enough to get published in some kinds of professional literature. How is that possible? It is because I have found it helpful for me (personally, not just as some utilitarian tactic) to incorporate at least some elements of modern academic practices, standards, and values into my transcultural project. Not everyone appropriating archaeologists' narratives does that. Some completely disregard modern academic practices, standards, and values. So I might differ a little with Holtorf when he says there cannot be any difference between academic and amateur archaeologies. There does not necessarily have to be an exclusive differentiation, but there could be a high degree of differentiation – if key values of academic archaeology are not part of the culturally hybrid nature of an individual practitioner of an amateur, or alternative, archaeology.

Considering the "scientific" appearance of my work, some will naturally want to call it "pseudoscience". Although such name calling may give the name caller some emotional satisfaction and feeling of superiority, it sidesteps the real epistemological issues. Philosopher of science Larry LAUDAN (1983: 125) states,

> "If we would stand up and be counted on the side of reason, we ought to drop terms like 'pseudo-science' and 'unscientific' from our vocabulary; they are just hollow phrases which do only emotive work for us."

The topic of what kinds of people and ideas should be allowed entrance into professional archaeological forums is contentious. The trend over the past couple of decades has been toward inclusion and diversity, but there are also those who want to see more exclusivity—a uniform society of properly credentialed professional archaeologists, adhering to a single, positivist, objective "science" of archaeology, speaking with one voice. In Helsinki, during the discussion after the group of papers that included my own, a participant in the session, in a comment directed to me, pointed out that there are disagreements among mainstream archaeologists. Of course, this is true. Disagreements may take place among those who adhere to a single, positivist, objective "science" of archaeology, but only within explicitly and tacitly prescribed limits. Those who accept these limits place those who transgress them outside acceptable discourse in professional archeology.

Recently, archaeologist Cornelius Holtorf inspired controversy among professional archaeologists when he invited Semir Osmanagich, proponent of the Bosnian pyramids (which professional archaeologists believe to be nothing more than natural formations), to speak at the archaeology department of Linnaeus University in Sweden. Opinions for and against what Holtorf did are recorded in a forum section of *AP: Online Journal in Public Archaeology* (PRUITT *et al.* 2012), with some contributors to the forum taking some version of an exclusivist position and others taking some version of an inclusivist position. Taking a version of the exclusivist position was Geoff Carver, who thought it was not good that Holtorf invited Osmanagich to speak at his university. Carver said, "Before archaeologists can attain the professional status of heart surgeons or airline pilots, we need to be serious and present a more unified face to the public" (PRUITT *et al.* 2012: 44). Among those taking a more inclusivist position was Tera Pruitt, who outlined three approaches professional archaeologists can take toward alternative archaeologies and archaeologists:

> "One approach has been to scoff and rebuke. [...] A second approach that archaeologists have taken is to simply ignore the alternative archaeology and hope it goes away. [...] A third approach, which is the most complex and multifaceted, has been to engage with and study [them]" (PRUITT *et al.* 2012: 33–34).

For her, this does not mean agreeing with them. But Pruitt (PRUITT *et al.* 2012: 34) nevertheless asks,

> "[...] if we, as archaeologists, do not engage with alternative archaeologists – if we do not actively interact with them and understand why they do the things they do – how else can we understand why such projects accumulate such influence?"

I think an inclusivist engaged approach more accurately reflects today's transcultural realities than does an exclusivist approach. And many professional archeologists appear to agree. In my case, I certainly have not forced the academic review committees of the EAA and WAC to accept my abstracts. Nor have I forced the editors of professional publications to accept my contributions. Nor have I forced universities to allow me to lecture on their premises. I do, however, respect the right of professional archaeologists to decide who should have access to their conferences, educational institutions, and professional publications. If in the future, there is some universal agreement among professional archaeologists that I not be allowed access to their conferences, classrooms, and publications, then so be it.

My other publics

I will now switch back to my provisional identification as a mediator, which is the main identification I adopt in this report. I have just discussed one of my publics, the community of professional archaeologists. And because I am giving this report to professional archaeologists, I have devoted most of it to my dealings with that public. But I also transmit narratives I have appropriated from archaeologists to other publics. One such public is "the general public." For this public, I produced the book *The Hidden History of the Human Race* (CREMO – THOMPSON 1994a), an abridged popular edition of my book *Forbidden Archeology* (which was intended primarily for my academic public). This abridged popular edition has been published in sixteen languages.[3]

---

[3] French (CREMO – THOMPSON 2002a), Spanish (CREMO – THOMPSON 1999a), Portuguese (CREMO – THOMPSON 2004), Italian (CREMO – THOMPSON 1997b), Russian (CREMO – THOMPSON 1999b), Bulgarian (CREMO – THOMPSON 2002b), Czech (CREMO – THOMPSON 2001a), Polish (CREMO – THOMPSON 1998b) Estonian (CREMO – THOMPSON 2006a), Croatian (CREMO – THOMPSON 2005), Slovenian (CREMO – THOMPSON 2001b), German (CREMO – THOMPSON 1994b), Danish (CREMO – THOMPSON 2006b), Hungarian (CREMO – THOMPSON 1997a), Japanese (CREMO – THOMPSON 1998a), and Chinese (CREMO – THOMPSON 2012). The full unabridged edition of *Forbidden Archeology* (over 900 pages), has been published in German (CREMO – THOMPSON 2006c) and Romanian (CREMO – THOMPSON 2008).

I also communicate with the general public through the media, including newspapers, radio, television, and the web. Factually, the general public is segmented. For example, there are segments of the public interested in various categories of "alternative science". I communicate with these publics through a regular column that I write for *Atlantis Rising* magazine. The column is titled "The Forbidden Archaeologist" and I have collected many of the columns in a recent book (CREMO 2010). Another public for me is my coreligionists in India and elsewhere. My basic strategy is to reach as many publics as possible by as many means of communication as possible. Not all my communications with these publics are initiated by me. There are many mediators who independently appropriate my narratives, and transmit them to these many publics. Therefore, just as I appropriate archeologists' narratives, many people also appropriate mine.

The value of my appropriations

In the concluding part of this report, I return to a point made in the session abstract. Some archaeologists are unhappy with the quality of public appropriations of archaeologists' narratives. To them, Meier and Niklasson (MEIER – NIKLASSON 2012: 73) say:

> "We regard such public receptions of archaeological narratives as productive transformations in their own right and reject an old-fashioned notion of academic knowledge versus the misunderstood and deteriorated narratives of 'the villagers'."

I am not suggesting that Meier and Niklasson believe there are no cases in which academic knowledge has been misunderstood and deteriorated. Neither am I suggesting they would necessarily include my transformations of archaeological narratives among those that they consider "productive". I have noticed, however, that some professional academics do seem to see some traces of value in my appropriations of archaeological narratives. I realize that these academics are not giving unqualified endorsements of my work, and I do not intend to give that impression. I am merely pointing out that embedded in their analyses and critiques of my work there are some acknowledgments of "productive" aspects.

In a lengthy review article about my work, Jo Wodak and David Oldroyd (WODAK – OLDROYD 1995: 207) asked, "So has *Forbidden Archeology* made any contribution at all to the literature on palaeoanthropology?" They concluded,

"Our answer is a guarded 'yes', for two reasons." First, "the historical material [...] has not been scrutinized in such detail before," and, second, the book does "raise a central problematic regarding the lack of certainty in scientific 'truth' claims."
Expanding on their first point, Wodak and Oldroyd wrote:

> "It must be acknowledged that *Forbidden Archeology* brings to attention many interesting issues that have not received much consideration from historians; and the authors' detailed examination of the early literature is certainly stimulating and raises questions of considerable interest, both historically and from the perspective of practitioners of SSK [sociology of scientific knowledge]. Indeed, they appear to have gone into some historical matters more deeply than any other writers of whom we have knowledge" (WODAK – OLDROYD 1995: 198).

This suggests that one productive aspect of my appropriation of archaeologists' narratives is that it helps keep the complete archive of archaeological data in view. Wodak and Oldroyd acknowledge that selective attention to only part of the entire archaeological record (the part that supports the currently dominant narrative) is a problem:

> "Certainly, some pre-*FA* [*Forbidden Archeology*] histories of palaeoanthropology, such as Peter Bowler's, say little about the kind of evidence adduced by C&T [Cremo and Thompson], and the same may be said of some texts published since 1993, such as Ian Tattersall's recent book" (WODAK – OLDROYD 1995: 197).

Archaeologists should not let the parts of the data set that do not support currently dominant narratives slip completely into oblivion. Data that does not make sense today may make sense in the light of future discoveries.

And now let us consider the epistemological point made by Wodak and Oldroyd, namely that my work addresses the problem of the certainty of scientific 'truth' claims. Let me give a more complete version of the passage I quoted above:

> "C&T [Cremo and Thompson] do raise a central problematic regarding the lack of certainty in scientific 'truth' claims, not so much because such claims are necessarily influenced or driven by vested interests but because the status of all knowledge is inherently a matter of degrees of probability and emerges as the result of social negotiations in concert with observation and inference. [...] when all that we can aspire to is a reduction in the level of doubt, recognition that this is the case is a healthy state of affairs, and those scientists who insist that evolution is a fact might be better advised to recast this as a 'highly probable theory'. [...] However, if C&T do succeed in showing that in palaeoanthropology, as in other sciences, knowledge is in part socially constructed, *FA* [*Forbidden Archeology*], their own massive intervention in palaeoanthropological debates, via its historiography, can equally be seen as a social act" (WODAK – OLDROYD 1995: 207).

I agree, and in this report, I have attempted to provide some insight into the nature of that social act.

Also addressing the epistemological point raised by Wodak and Oldroyd, archaeologist Marylène Pathou-Mathis wrote in *L'Anthropologie*:

> "M. Cremo and R. Thompson have willfully written a provocative work that raises the problem of the influence of the dominant ideas of a time period on scientific research. These ideas can compel the researchers to orient their analyses according to the conceptions that are permitted by the scientific community" (PATHOU-MATHIS 1995: 159; translated from the French original).

With all this in mind, I believe that professional academic archaeologists may take my appropriations of their narratives as "productive", to the extent that they contribute to:

(1) a better understanding of the epistemology of archaeology,
(2) a better understanding of the history of archaeology,
(3) a better understanding of the complete set of archeological data, and
(4) a better understanding of the need for archaeology to remain open to multiple narratives about human origins and antiquity.

Alternatively, professional archaeologists may take my appropriations of their narratives as "misunderstood and deteriorated" – or perhaps as both "productive" and "misunderstood and deteriorated", to different degrees of each.

Some final thoughts on what I do and why I do it

In the Helsinki session, there was certainly some acknowledgment of the reality of unproductive, or even dangerous, appropriations of archaeological narratives. One must take into account the purposes for which archaeological narratives are being employed. Eszter BÁNFFY (this volume: 271) of Hungary, in the abstract of her paper, mentions a phenomenon she observed in her country: "the unexpected and aggressive spread of fringe archaeological ideas, often ultra-nationalistic." She expanded upon this point during her talk in Helsinki, pointing out that such nationalistic ideas incorporate material appropriated from mainstream archaeological sources. In the discussion that followed, several participants in the session voiced their concern about this phenomenon: "You have frightened me," one participant said. Of course, it is not only fringe groups that can use archaeology for nationalist purposes. The eminent historian of archaeology Bruce Trigger noted,

> "Most archaeological traditions are probably nationalistic in origin. The development of European prehistoric archaeology was greatly encouraged by the post-Napoleonic upsurge of nationalism and romanticism. Some of this archaeological activity was directed towards strengthening patriotic sentiments and in these cases it often received substantial government patronage" (TRIGGER 1984: 358).

Those patriotic sentiments contributed to massive wars and other manifestations of violence in the nineteenth and twentieth centuries.

However, my kind of archaeology is not motivated by nationalism, including Indian nationalism. My study of the spiritual texts of ancient India, like the *Bhagavad-gita*, leads me to the conclusion that the conscious self, the *atma*, has an identity separate from that of the material body. The *atma* is the real self. One should not therefore overly identify with the designations of the body, such as race or nationality, and engage in conflict, or exploitative behavior, based on these superficial designations. The *Bhagavad-gita* (5.18) says *panditah sama darshinah*: a pandit, or wise person, sees all living beings as equal. I do not claim a monopoly on religious truth. I believe similar insights can be found in all authentic spiritual traditions, including Christianity, Judaism, Islam, and various tribal worldviews. The public purposes for which I appropriate archaeological narratives ultimately involve moving humankind in the direction of transcending conflict based on inappropriate identification with nation, race, or religion.

Although it would have been possible for me to carry out my work in isolation from the community of professional archaeologists, I have chosen to carry it out in such a way that I am in dialog with that community. In my opinion, that is good for both parties.

References

BÁNFFY, ESZTER (this volume), Disarmed post-socialist archaeologies? Social attitudes to interpreting the past – an interim report from Hungary. This volume 271–293.

CREMO, MICHAEL A. (1998), *Forbidden archeology's impact.* Los Angeles.

CREMO, MICHAEL A. (1999), Puranic time and the archaeological record. In: Tim Murray (ed.), *Time and archaeology.* London. 38–48.

CREMO, MICHAEL A. (2002), The later discoveries of Boucher de Perthes at Moulin Quignon and their impact on the Moulin Quignon jaw controversy. In: Goulven Laurent (ed.), *Proceedings of the XXth International Congress of History of Science (Liege, 20-26 July 1997), Volume X, Earth Sciences, Geography and Cartography.* Turnhout. 39–56.

CREMO, MICHAEL A. (2004), *Human devolution: A Vedic alternative to Darwin's theory.* Los Angeles.

CREMO, MICHAEL A. (2008), Excavating the eternal: an indigenous archaeological tradition in India. *Antiquity* 82, 178–188.

CREMO, MICHAEL A. (2009), The discoveries of Carlos Ribeiro: a controversial episode in nineteenth-century European archeology. *Journal of Iberian Archaeology* 12, 69–89.

CREMO, MICHAEL A. (2010), *The forbidden archeologist.* Los Angeles.

CREMO, MICHAEL A. (2012a), *My science, my religion: Academic papers 1994-2009.* Los Angeles.

CREMO, MICHAEL A. (2012b), An insider's view of an alternative archaeology. In: Anna Simandiraki-Grimshaw – Eleni Stefanou (eds), *From archaeology to archaeologies: The 'other' past.* BAR International Series 2409. Oxford. 12–17.

CREMO, MICHAEL A. – RICHARD L. THOMPSON (1993), *Forbidden archeology.* Los Angeles.

CREMO, MICHAEL A. – RICHARD L. THOMPSON (1994a), *The hidden history of the human race.* San Diego.

CREMO, MICHAEL A. – RICHARD L. THOMPSON (1994b), *Verbotene Archäologie.* Essen.

CREMO, MICHAEL A. – RICHARD L. THOMPSON (1997a), *Az emberi faj rejtélyes eredete.* Budapest.

CREMO, MICHAEL A. – RICHARD L. THOMPSON (1997b), *Archeologia proibita: la storia segreta della razza umana.* Bresso.

CREMO, MICHAEL A. – RICHARD L. THOMPSON (1998a), *『人類の隠された起源』* [*The hidden history of the human race*]. Tokyo.

CREMO, MICHAEL A. – RICHARD L. THOMPSON (1998b), *Zakazana Archeologia: Ukrtya Historia Człowieka.* Wroclaw.

CREMO, MICHAEL A. – RICHARD L. THOMPSON (1999a), *La historia oculta de la especie humana.* Mexico.

CREMO, MICHAEL A. – RICHARD L. THOMPSON (1999b), *Неизвестная история человечества.* Moscow.

CREMO, MICHAEL A. – RICHARD L. THOMPSON (2001a), *Skrytá historie lidstva.* Praha.

CREMO, MICHAEL A. – RICHARD L. THOMPSON (2001b), *Prikrita zgodovina človeške rase.* Ljubljana.

CREMO, MICHAEL A. – RICHARD L. THOMPSON (2002a), *L'Histoire secrete de l'espèce humaine.* Monaco.

CREMO, MICHAEL A. – RICHARD L. THOMPSON (2002b), *Тайната история на човешката цивилизация.* Sofia.

CREMO, MICHAEL A. – RICHARD L. THOMPSON (2004), *A história secreta da raça humana.* São Paulo.

CREMO, MICHAEL A. – RICHARD L. THOMPSON (2005), *Skrivena povijest ljudske vrste.* Zagreb.

CREMO, MICHAEL A. – RICHARD L. THOMPSON (2006a), *Inimkonna varjatud ajalugu.* Tallinn.

CREMO, MICHAEL A. – RICHARD L. THOMPSON (2006b), *Forbudt Arkæologi.* Mørkøv.

CREMO, MICHAEL A. – RICHARD L. THOMPSON (2006c), *Verbotene Archäologie.* Rottenburg.

CREMO, MICHAEL A. – RICHARD L. THOMPSON (2008), *Arheologia interzisă.* Constanța.

CREMO, MICHAEL A. – RICHARD L. THOMPSON (2012), *被掩盖的人类历史*[*The Hidden History of the Human Race*]. Shandong (Peoples Republic of China).

HOLTORF, CORNELIUS (2006), One world archaeology today. *Archaeologies: Journal of the World Archaeological Congress* 2/2, 87–93.

LAUDAN, LARRY (1983), The demise of the demarcation problem. In: Robert S. Cohen – Larry Laudan (eds), *Physics, philosophy, and psychoanalysis.* Dordrecht. 111–127.

LEAKEY, MARY D. (1979), Footprints in the ashes of time. *National Geographic* 155, 446–457.

LOCKE, SIMON (1999), Review: Forbidden archeology's impact. *Public Understanding of Science* 8/1, 68–69.

MARKS, JONATHAN (1994), Review: Forbidden archeology: The hidden history of the human race by Michael A. Cremo and Richard L. Thompson. *American Journal of Physical Anthropology* 93, 140–141.

MEIER, THOMAS – ELISABETH NIKLASSON (2012), Session A16: Public appropriations of archaeologists' narratives. In: Eeva-Maria Viitanen (ed.), *Abstracts, 18th Annual Meeting, European Association of Archaeologists.* Helsinki. 73–74.

MURRAY, TIM (1995), Review: Forbidden Archeology. *British Journal for the History of Science* 28, 377–379.

PATHOU-MATHIS, MARYLÈNE (1995), Forbidden archeology (book review), *L'Anthropologie* 99/1, 159.

PRUITT, TERA – JOHAN NORMARK – LORNA RICHARDSON – BEATRIZ COMENDADOR REY – GEOFF CARVER – CORNELIUS HOLTORF – JACOB HILTON (2012), Forum: The limits of collaboration. Osmanagić on the campus. *AP: Online Journal in Public Archaeology* 2, 24–54. (www.arqueologiapublica.es [June 17, 2013]) (Each of the first five authors made an individual contribution to the forum, and Holtorf and Hilton coauthored their contribution).

RIBEIRO, CARLOS (1873), Sur des silex taillés, découverts dans les terrains miocène du Portugal. *Congrès International d'Anthropologie et d'Archéologie Préhistoriques, Bruxelles 1872, Compte Rendu,* 95–100.

RUTOT, AIMÉ (1907), Un grave problem: une industrie humaine datant de l'époque oligocène. Comparison des outils avec ceux des Tasmaniens actuels. *Bulletin de la Société Belge de Géologie de Paléontologie et d'Hydrologie* 21, 439–482.

SIMANDIRAKI-GRIMSHAW, ANNA – ELENI STEFANOU (eds) (2012), *From archaeology to archaeologies: The 'other' past.* BAR International Series 2409. Oxford.

TRIGGER, BRUCE G. (1984), Alternative archaeologies: nationalist, colonialist, imperialist. *Man (New Series)* 19/3, 355–370.

WELSCH, WOLFGANG (1999), Transculturality – the puzzling form of cultures today. In: Mike Featherstone – Scott Last (eds), *Spaces of culture: City, nation, world.* London. 194–213.

WHITNEY, JOSIAH D. (1880), The auriferous gravels of the Sierra Nevada of California. *Harvard University, Museum of Comparative Zoology Memoir* 6/1.

WODAK, JO – DAVID OLDROYD (1995), 'Vedic creationism': a new twist to the evolution debate. *Social Studies of Science* 28, 192–213.

“Selecting what is important for the reader”: about appropriations and transformations of archaeology in the mass media

DIANE SCHERZLER

Abstract:
Archaeologists do not think highly of journalists and the mass media, it is true. Mistakes, omissions, exaggerations and clichés which scientists discover in the media are often seized upon (see ASCHERSON *2004: 145ff., 154;* BRITTAIN – CLACK *2007: 17, 23;* KIRCHER *2012: 259ff.). They complain that the way archaeologists are portrayed in the media differs from the discipline's own self-perception and that "the public perception is formed almost exclusively by a media-staged and distorted image of archaeology" (*FELDER et al. *200: 163). This article would like to show the extent to which criteria other than scientific standards are (also) important in the assessment of journalistic quality. How and why do the media transform the portrayal of archaeology? The aim is not to apportion blame: journalism is undoubtedly a flawed system, and not every newspaper article, film or radio report provides perfect quality by any means. Several examples will serve to show how archaeologists as well can influence the scale of mistakes or the quality of a journalistic report. My perspective is that of an archaeologist who has been able to acquire many years of experience in dealing with journalists and archaeologists. From 2002 to 2012 I worked as an editor and author at one of the largest public media companies in Germany and regularly dealt with archaeological issues. I teach science communication at several universities, write articles like this and hold media training courses for scientists who want to improve their collaboration with the media. Over the years I have worked with some 450 archaeologists from different countries and specialist fields and have become familiar with their opinions. Their views of the media are incorporated into this chapter together with my experience as a science journalist.*

What must a “good” journalistic report have? Whereas scientists would transfer the scientific community’s own criteria and would possibly say that the factual correctness and completeness of all scientifically important aspects were absolutely imperative, journalists have completely different criteria and constraints. A couple of examples:

The top priority for a journalist is that the reader, listener or viewer can understand a report. If they cannot, there is a high probability that the reader will put the book down or that the listener or viewer will switch over to another channel within seconds. Scientists must express themselves clearly in order for the report overall to be intelligible. But what is a journalist to do in an interview when an archaeologist produces the following sentence: "We found secondary bone deposits everywhere"? To a non-archaeologist this is completely unintelligible without further explanations, and when the scientist cannot make themself understood when asked to explain, it is down to the reporter to provide the explanation. It is extremely likely that this is flawed. What about the statement: "Many burials were met in the church" (German: "*In der Kirche wurden zahlreiche Bestattungen angetroffen*")? Although this is not a specialist expression, it is a typical example of jargon in German scientific language, which is even unintentionally funny: I hope no-one has ever met a burial! Archaeologists have used both these turns of phrase in interviews with me. Since they were not aware that I am a prehistorian,[1] I know that they would have expressed themselves in this way to any journalist. This means they cannot be quoted, however, and so a journalist is not going to interview them in the short or long term: the effort is simply too great. While archaeologists may think from their point of view that understanding the discipline means knowing many scientific facts, theories or methods, for journalists being understood means that the reader understands a fact when they read an article for the first time. This applies to television and radio even more: if one excludes the growing use of media libraries, which allow people to view and listen to reports again and again, explaining in television and radio means that the viewers and listeners have to understand what is said immediately or with a few seconds delay at most, otherwise it becomes impossible for them to follow the film or the radio broadcast any longer. Thus "understanding archaeology" often means something completely different for archaeologists and journalists and each has good reasons for thinking as they do. The two parties are rarely aware of this when they meet, however.

Television places its own particular demands on a topic: TV must have pictures to show. Time and again archaeologists notice that some tasks such as the excavation are shown very frequently on television, while other tasks – this includes literature searches – are shown only very seldom (BENZ – LIEDMEIER

1 In interviews with archaeologists I do not explicitly point out that I studied prehistory. My experience has shown that this usually leads to my interview partner preferring to conduct an academic discussion with me or treating me like a student. Apart from this not being an appropriate role allocation, it would be almost impossible to compile a comprehensible, short report from this type of conversation in the short time available.

2007: 170; HOLTORF 2007: 86). Television author Christian Frey, who has been making documentaries on the history of the 20th century for many years, explains why this is so in his medium:

> "Everything that can be shown in pictures, which contains action, can be transported well on television. This is much more interesting for the viewer [...] than a scientist studying literature, for example. Just imagine a film about studying the sources of the New Testament. If you were only to show someone who is bent over the Qumran scrolls all the time, then quite simply nobody would watch it" (pers. comm. 28 August 2009).

Frey adds: "The more tangible a story – the more it is a story with a protagonist and a conflict, with a beginning, a progression and an end – the better it can be represented on television (pers. comm. 28 August 2009)." A further significant factor: television is by far the most complex and therefore the most expensive mass medium to produce. The television author Tamara Spitzing, herself a trained archaeologist, describes the consequence:

> "The editors naturally want to reach a large audience. They *have to* reach a large audience. That is to say: there have to be – at least at peak broadcasting times – opulent pictures and theatrical scenes. And the story has to be exciting. Topics *must* be easy to understand and phrased in a greatly exaggerated way, sometimes even sensational" (SPITZING 2010: 221).

In the medium of television the dependence on images means that not all archaeological topics and not all aspects of an archaeological topic can be presented. At the same time television emphasises visually attractive aspects more strongly and even stages them, with the aid of re-enactments, for example.

The broadcasting time in radio and television is very limited, the space in a newspaper a valuable commodity. At the same time, there is always a surplus of possible topics with which press information officers approach journalists. Which archaeological topics are interesting for editors and which are not? Journalists put a topic into a broader, non-scientific context which is relevant and of interest to the audience in question. The items the media communicate must therefore satisfy certain criteria, so-called news factors. It is important for journalists that an issue concerns a large number of people, for example, that it is new and that it is set in the spatial or social vicinity of the addressee (RUHRMANN – GÖBBEL 2007), for example. Cornelia Varwig, editor for archaeological issues in the German science journal "*bild der wissenschaft*" writes:

> "It is possible to distinguish at least seven crucial factors for the selection of archaeological topics in '*bild der wissenschaft*': Novelty, prominence, spatial and/or cultural vicinity, fascination, mystery/secret, superlative, exotics" (VARWIG 2010: 69).

Tamara Spitzing adds:

> "Topics which I find interesting as a journalist must tell a story or continue a story in addition to the purely scientific facts. The potential viewer must be able to put the topic into a context which they already know" (pers. comm. 4 August 2009).

There are several other news factors, but these should not be imagined as a checklist for the editor, more as something which constitutes their "gut feeling" or "good nose" for or against a topic. These factors carry different weightings, depending on the medium and the audience. The more news factors a story can combine, the more probable it is that journalists will take it up. Incidentally: scientific relevance is also one of these factors, but is only one among many others (see also LEHMKUHL 2006: 102).

Whereas scientists are often highly specialised and spend months and years investigating something, journalists must in most cases be generalists so that they can be used for a wide range of topics. The local reporter reports today on an archaeological dig, but tomorrow on a fire caused by an electrical fault in the town hall, and the day after on the nature reserve on the outskirts of the city. A science journalist deals with archaeology as well as with geology and linguistics. And even freelancers who have specialised in archaeology and offer their reports to several media companies deal with topics such as the graveyard from Merovingian times and also with Neanderthal man, palaeogenetics or the handling of stolen antique artefacts. The renowned author and science journalist Gero von Randow gave a talk in which he put it like this:

> "It is a well known saying: the problem with the facts is that there are so many of them. Science journalists have at best an extensive knowledge of one field, maybe some knowledge of four or five fields, and otherwise they have contacts. It cannot be any other way" (VON RANDOW 2000: 173).

Articles, films and radio reports are nearly always produced under enormous pressures of time, and deadlines are often stipulated to within a minute. Alexander Mäder is chief science editor of the *Stuttgarter Zeitung*, a respected newspaper for Stuttgart and the surrounding area. He allows his authors two to three days time to prepare, "if I believe that the topic justifies it in order to build up competence in this field in the editorial department" (pers. comm. 18 January 2011). Often, much less time is available, however. Mäder says: "There are other texts which we pull out of a hat, so to speak: news of a study lands on our desk, we quickly make a phone call, and then we immediately start to write" (pers. comm. 18 January 2011). Mäder has intensive connections with other science journalists and knows the Germany newspaper landscape very well.

> "I believe" he says "that many short news reports in German daily newspapers are researched and written by the author in the news agency in one, maybe two hours. When we then receive the report from the agency, it is revised within ten, fifteen minutes and given a title. We have three or four of those short news reports in the science section every day" (pers. comm. 18 January 2011).

Even journalists think that this narrow time frame is highly problematic. Journalism has been fighting for years against staffing cuts in editorial departments. The reality is: only a small number of the journalists who work with archaeologists can really have a high degree of familiarity with the topic; all journalists have very little time to produce and small amounts of text or broadcasting time available.

One of the most frequent complaints from scientists of all disciplines about the media is that they are full of mistakes. This is correct: they are full of mistakes, and many journalists will agree with this (e.g. SILVERMAN 2009). If one looks at the "mistakes" in detail, however, quite a few of them turn out to be simplifications and omissions. Both are indispensable in journalism – nobody wants to be informed about all the facts of this world in their entire complexity and all details, or would be able to grasp them. Although simplifications and omissions can lead to factual errors, this does not have to be the case by any means. Judith Rauch, former editor of "*bild der wissenschaft*" says: "We will never be able to be complete. After all, journalism consists in selecting what is important for the reader. Never completeness!" (pers. comm. 16 September 2009). In contrast to science, the completeness of information can therefore not be a quality criterion in journalism (see SCHERZLER 2012: 79)! In one of my media training courses a young archaeologist complained about the incorrectness of the media: a Palaeolithic figurative image which was the subject of an article in the local newspaper could have been interpreted in ten different ways. He had listed them all to the reporter in the interview. The reporter then mentioned only three of them in his text; the archaeologist felt this was a clear misrepresentation. The article had about 400 words, and the archaeologist also conceded that they had known the length of the planned article even before the interview. It is impossible to list and explain all ten possible interpretations in 400 words. In addition, the average reader of a local newspaper does not really want to know about all interpretative approaches, but would probably like to know only the most important ones. From a journalistic point of view it was completely correct to limit the description for the reasons described above. An archaeologist experienced in science communication would have ensured in the interview that they themselves named the most important interpretative approaches and did not leave the selection to the journalist.

Some archaeologists think it would be very easy for journalists to produce good, correct reports. They need only adhere more closely to the information provided by the scientists. William E. BOYD (1995: 52) considers successful newspaper reports and radio interviews to be those which adhere very closely to his press releases, sometimes word-for-word, and which tackled the issues which he himself considered to be the most important. Metin YEŞILYURT (2007: 61) claims that there "will be hardly any differences in opinion between scientists and science journalists regarding portrayal in the media", because a science journalist would obviously make reference to the science they are reporting on. So Yeşilyurt's view is that good science journalists report in a way which pleases the archaeologist. In media training courses scientists often ask me: "Why couldn't the journalist simply write what I dictated to them? It would then have been correct!" Here at the latest a fundamental misunderstanding of the role and function of journalism becomes clear: quality journalism has to be guided by its audience, and not by whether it always pleases those about whom it reports.

The audience wants to be able to form an opinion with the aid of journalism, or to judge a fact. The fact that journalism must therefore not only reproduce what scientists, politicians or industry bosses would like hopefully becomes clear. Most journalists do not consider themselves as intermediaries who must ensure the acceptance of archaeology and the dissemination of archaeological knowledge in society, or even as the scientists' "translators", but much rather as representatives of the public (SCHERZLER 2012: 80). If journalists now fulfil the wish of some archaeologists to keep closely to their press releases or statements, then this is not a careful course of action from a media point of view, but in contrast rather more a lack of journalistic quality. The communication scientist Matthias Kohring writes: "The trustworthiness of journalism depends directly on the distance it keeps from the object of its reporting (this distinguishes it fundamentally from PR)" (KOHRING 2004: 177). Science journalism is journalism which observes science and uses it as a source of information. It is not journalism which is scientific or carried out by scientists. Neither is it simply the dissemination of science.

And even if it wanted to: journalism cannot represent archaeology as it sees itself. Rather one can say that a transformation is achieved via the media. I have already written about the limitations of television earlier. In general, the media are very often event controlled. Something happens and journalists report about it. Science, in contrast, is a process: often slow, sometimes monotonous, full of errors, small successes and great doubt. Classical mass media can usually not represent this process, they overemphasise the events and act as if breakthroughs happened all the time. The fact that the reason for their reporting – a published paper, say – is not itself a breakthrough, but the result of something about which the media report very little or not at all, is part of the distortions by the mass media (cf. BENZ – LIEDMEIER 2007: 170). Journalists do not always understand the character of a scientific discourse and sometimes deem such a "dispute" between scientists as a sign of great conflict and thus worthy of reporting.

Until now I have described what good journalism is *not* or does *not* have to be: a complete and perfect representation of facts, if possible word-for-word, covering all aspects of the topic to the satisfaction of the source (i.e. the scientist). But what *is* good journalism? This is obviously not so easy to define, but it is imperative to inform truthfully,[2] as neutrally as possible and according to the best of one's knowledge and belief. Principles of good journalistic practice are accuracy (but

2 In this text "truth" means not to lie, not to falsify, not to uncritically copy one's personal opinion without making this clear.

not in the sense of completeness, rather in the sense of correctness), in addition to balance and independence (cf. THOMSON REUTERS CORPORATION 2008: 3). Topicality and relevance of a topic are also crucial for journalistic quality, as are careful research and critical assessment of sources. Another measure for the quality of journalism is whether the public can get an idea of archaeology with the aid of a journalistic report, regardless of how archaeology presents itself (cf. KOHRING 2004: 178).

As I said earlier: archaeologists are not happy that they are stereotypically presented in the media as excavators, discoverers and treasure hunters. On the one hand their annoyance is understandable. My view, however, is that they fail to see how strongly they themselves communicate: "we have found" instead of "we have gained an insight". Without having investigated this systematically, my impression from many years of collaboration with archaeologists is that when communicating outside the discipline they have quite a strong fixation with excavations and artefacts. Journalists experience this in press releases, on-site appointments at digs or exhibition openings, and in interviews. Journalists thus naturally learn to have a very specific view of archaeology. Time and again I have read the phrases "sensation", "breakthrough" or "early history must be rewritten" in press releases of archaeologists as well. And all this must naturally be seen in interaction with the presumed expectations of media and public, which archaeologists would like to fulfil in order to get into the headlines. Tamara Spitzing voices her criticism of this:

> "Archaeology as a science has allowed how it presents itself to be strongly commercialised as well. Archaeological events are sometimes presented as if they were pop concerts. Some archaeologists also emphasise the sensational aspects of their research when they deal with the media. They know that this is an interactive process. It is therefore not just the journalists, but also some archaeologists who emphasise the sensational aspects of this discipline" (pers. comm. 4 August 2009).

During my ten years as a journalist no archaeologist ever seized the opportunity to make me aware of their own accord that it is not only the find but the findings in particularly which are of considerable importance. Nobody has ever told me that a bone fragment can provide more information than the most wonderful brooch. I ask myself how is a journalist with no archaeological training ever going to hear about it? How can this thought ever occur to a journalist, who writes about a wide variety of topics on a daily basis?

I hope I have succeeded in illustrating that journalism and science sometimes have very different criteria for quality, and that the standards of science must not be applied to journalism without further thought. Journalists consider themselves as scouts of the audience in the world of science, but not as disseminators or even "translators" of research. Their professional distance to archaeology is not a deficit, but a sign of quality according to the standards of communication science. The media do not provide a perfect image of archaeology. They do not consider this to be their task, and neither are they in any position to do this. In contrast, their portrayal transforms archaeology. This paper did not aim to deal with communication deficits of archaeologists and weaknesses of journalism, even if they have been stated. Its main intention was to illustrate the differences which exist between the two systems, because it is then that potentials which go beyond mutual projections and accusations become visible, and these potentials allow archaeological topics to be portrayed in a better way in the mass media.

References

ASCHERSON, Neal (2004), Archaeology and the British media. In: Nick Merriman (ed.), *Public archaeology*. London. 145–158.

BENZ, MARION – ANNA KATRIEN LIEDMEIER (2007), Archaeology and the German press. In: Timothy Clack – Marcus Brittain (eds), *Archaeology and the media.* Walnut Creek. 153–173.

BOYD, WILLIAM EDGAR (1995), Media coverage of an archaeological issue: Lessons from the press release of initial radiocarbon dating results of a possible pre-Cook European ship at Suffolk Park, northern New South Wales. *Australian Archaeology* 40, 50–55.

BRITTAIN, MARCUS – TIMOTHY CLACK (2007), Introduction: Archaeology and the media. In: Timothy Clack – Marcus Brittain (eds), *Archaeology and the media.* Walnut Creek, CA. 11–65.

FELDER, KATHRIN – ISABELLA HAMMER – JULIANE LIPPOK – MAREIKE WULF (2003), Erkenntnisgewinn und Unterhaltungswert. Eine Analyse von Archäologiebildern in den Unterhaltungsmedien. *Ethnographisch-Archäologische Zeitschrift* 44, 161–182.

HOLTORF, CORNELIUS (2007), An archaeological fashion show. How archaeologists dress and how they are portrayed in the media. In: Timothy Clack – Marcus Brittain (eds), *Archaeology and the media.* Walnut Creek. 69–88.

KIRCHER, MARCO (2012), *Wa(h)re Archäologie. Die Medialisierung archäologischen Wissens im Spannungsfeld von Wissenschaft und Öffentlichkeit.* Bielefeld.

KOHRING, MATTHIAS (2004), Die Wissenschaft des Wissenschaftsjournalismus. Eine Forschungskritik und ein Alternativvorschlag. In: Christian Müller (ed.), *SciencePop. Wissenschaftsjournalismus zwischen PR und Forschungskritik.* Graz. 161–183.

LEHMKUHL, MARKUS (2006), Auswahlkriterien für Wissenschaftsnachrichten. In: Winfried Göpfert (ed.), *Wissenschafts-Journalismus. Ein Handbuch für Ausbildung und Praxis.* Berlin. 98–104.

RUHRMANN, GEORG – ROLAND GÖBBEL (2007), Veränderung der Nachrichtenfaktoren und Auswirkungen auf die journalistische Praxis in Deutschland. Concluding report for netzwerk recherche e.V. (http://www.netzwerkrecherche.de/docs/ruhrmann-goebbel-veraenderung-der-nachrichtenfaktoren.pdf [03.03.2013]).

SCHERZLER, DIANE (2012), A look in the mirror and the perspective of others. On the portrayal of archaeology in the mass media. In: Anna Simandiraki-Grimshaw – Eleni Stefanou (eds), *From archaeology to archaeologies – The "other" past.* British Archaeological Reports International Series 2409. Oxford. 77–85.

SILVERMAN, CRAIG (2009), *Regret the error. How media mistakes pollute the press and imperil free speech.* New York.

SPITZING, TAMARA (2010), Das Indiana Jones-Syndrom: Wissenschaftsfernsehen zwischen Wissenschaft und Kommerz. In: Hans-Joachim Gehrke – Miriam Sénécheau (eds), *Geschichte, Archäologie, Öffentlichkeit. Für einen neuen Dialog zwischen Wissenschaft und Medien. Standpunkte aus Forschung und Praxis.* Bielefeld. 219–225.

THOMSON REUTERS CORPORATION (2008), *Handbook of Reuters Journalism. A guide to standards, style and operations* (http://handbook.reuters.com/index.php [06.03.2013]).

VARWIG, CORNELIA (2010), Wie man Ausgräber angräbt. Über das Verhältnis von Journalismus und Archäologie. In: Hans-Joachim Gehrke – Miriam Sénécheau (eds), *Geschichte, Archäologie, Öffentlichkeit. Für einen neuen Dialog zwischen Wissenschaft und Medien. Standpunkte aus Forschung und Praxis*. Bielefeld. 161–180.

VON RANDOW, GERO (2000), *Populäre Wissenschaft – wer will das?* Speech at the summer fair of the Engineering Faculty at the University of Kiel, 30.6.2000 (http://www.tf.uni-kiel.de/dekanat-deutsch/berichte/rede-vonrandow.html [31.03.2009]).

YEŞILYURT, METIN (2007), Eine Stellungnahme zur Darstellung des "Archäologischen" in den Medien. Formen der Zusammenarbeit von Wissenschaft und Wissenschaftsjournalismus, Verweise auf allgemein gültige Normen journalistischer Ethik, Geschäfte und was in der Archäologie noch zu tun ist. *Archäologische Informationen* 30/2, 59–66.

Archaeology and identity in a Bavarian village – academic and local (hi)stories

THOMAS MEIER

Abstract:
In village chronicles experts are frequently asked to give their academic perspectives on the prehistory of the place. This paper starts with a discussion of how such a contribution might look like under the ethics of post-modernity. I suggest that we regard villagers (hi) stories as appropriations in their own right and side by side with academic versions. For example such academic and local (hi)stories are compared for three sites in the territory of a Bavarian village– with remarkable differences. Based on these examples mediators, modes and conditions are reflected on a more general level, which structure the appropriation of academic (hi)stories in a local public. While the category of the mediator becomes blurred between research and mediation, even mediated archaeological (hi)stories are not fully appropriated, but selected at a local level. The selection of (hi) stories on the one hand depends on the ways of making historical truth – with socially constructed authorities like school and academics being in an advantage to gain discursive power. On the other hand the narratives of discursive authorities have to meet further conditions to become appropriated locally. They need to adhere to physical places and they need to meet the general educational background, i.e. established master-narratives of "Great" history (Celts, Romans, Middle Ages, sacrificial site, watchtower).

As a local friend of mine once said, the village of Götting is unspectacular in almost every respect. It is situated in Upper Bavaria about 50 km southeast of Munich in the Mangfall valley and close to the rim of the Alps. It is here where we lay our scene for a case study on a local (non-)interaction between academic and public[1] appropriations of the past.[2]

The choice of Götting for this case study is purely personal coincidence: I came across the village for two reasons, both not related to this research: firstly,

---

1 As in this case the public I have chosen for my research is a very local village community, I am using the terms "public" and "local" quite synonymously throughout this paper. Therefore my term "local knowledge" bears hardly any reference to the local knowledge as Clifford GEERTZ (1983) has conceptualized it.

2 An earlier version of this paper was published in 2008. For this volume I have fully re-structured, focused and translated it.

fifteen years ago I started excavating a small priory about twenty kilometres east of Götting, which quickly developed into a project on human-environmental interaction and comprised a wider survey of the former monastic holdings including – among others – the lower Mangfall valley and Götting (MEIER 2004a; MEIER – SCHMITT 2007). Secondly, among German archaeologists, at least those interested in the Migration Period, the village of Götting is famous for a female burial of the years around 400 AD (KELLER 1971). When asked for a contribution to the *Festschrift* of my academic teacher Max Martin in 2002 I decided to present to him a re-evaluation of this burial (MEIER 2004b). As an unintended consequence of that publication I was asked to contribute "something" on the pre- and protohistory of Götting to a chronicle of the village some years later in 2007 (MEIER 2008). It was at that moment, when I started to reflect on my perspective on my and "their" – the villagers' – archaeologies and what I assumed to be "their" perspective.

In post-modernity academic history has become multiple, it has lost its innocent claim to tell the one and only true story. Historical narratives[3] today no longer speak of truths, but of realities, perspectives and opinions of different people and groups. This is not only a matter of multi-perspectivity in the past. In our days, too, history has become manifold, there is no longer a joint history of "the" society, but the multiple and different perspectives of social groups on their pasts as for example the papers of this volume show. In itself the existence of such multiple and perspective pasts is far from new, but with post-modern ethics of plurality the plurality of (hi)stories becomes reflected and is legitimized. Non-academic perspectives on the past are no longer considered as false or fringe, but as valuable constructions in their own right and for their own groups and purposes. Their legitimacy cannot be judged against an alleged historical truth (in this point I do not agree with BINTLIFF this volume: 246) as such a truth itself is dismantled as a discursive construction of authority (FOUCAULT 1971: 15–23; 1977; on the socio-historical conditions of truth SHAPIN 1995). The only yardstick against which the acceptability of a perspective and story can be measured is

[3] On the narratological character of history cf. WHITE (1973; 1987); for contradiction NÜNNING (1999); for a survey MEUTER (2011). For archaeology cf. PLUCIENNIK (1999; 2010) and the summary by VEIT (2010).

an ethical one: Whether it is respectful to and allows for other perspectives and narratives (WELSCH 1991).[4]

Such multiple (hi)stories regard different events and things as important and knit them together to different webs of (hi)stories, each of them being reasonable, conclusive and persuasive within their respective groups (cf. CONRAD – KESSEL 1994; GOETZ 2001).[5] If (hi)stories are negotiated within groups relative to their respective world-views, their concepts of reason and meaning, each of these (hi)stories is primarily meaningful to the respective group which has constructed it. In a plural, democratic society academic versions about the past are group-related narratives as well; the academic vision of the past is only one vision competing with others and fully unequipped with any heightened moral or argumentative powers (e.g. WELSCH 1991): Multiple in themselves academic narratives cannot necessarily claim to be more meaningful or important outside the academic sphere than any other story. An academic narrator selects the ingredients of her/his story according to the paradigms, measures of value and plausibilities of his peer group, s/he follows the rules of argumentation and the fashions of importance of her/his discipline and consequently her/his narrative is primarily meaningful to those sharing her/his academic perspective. As any other narrator s/he tells a story for her/his own social and intellectual group reaffirming her-/himself and her/his readers that they belong together because they are thinking in the same way and because they share the same view on the past. S/He founds a shared identity, a feeling of community fed by a shared history and tradition, which often enough constructs and legitimates the existence of the group (HOBSBAWM 1983).

In enlightenment ideology, however, academic narratives – originally in the wake of theology – were esteemed to be closer to an absolute truth due to their elaborate rules of rationality than any other narratives. They were equipped

---

4 Eszter Bánffy's paper (BÁNFFY this volume) shows the limits of post-modern plurality, if one perspective raises claims of absolute truth and tries to marginalize and rule out other perspectives. Such ethical limits at first sight seem to contradict the post-modern essence of radical plurality, but such an objection overlooks the fact that the post-modern constitution is based on an ethical concept with just this radical plurality at its core (WELSCH 1991: esp. 25, 28).

5 These conditions of post-modern narratives are often ignored by the blunt objective of "*anything goes*", discrediting post-modernity to be another word for arbitrariness. Post-modern narratives as any other narratives have to be consistently constructed according to rules of plausibility. So it is not the narratives, which are arbitrary, nor has the claim for consistency become disputable, but the specific sets of rules are negotiable and have to prove their meaningfulness in specific contexts.

with moral authority and legal power – not the least by making them the only narratives pupils were obliged to learn in school, which roughly simultaneously became compulsory. Compulsory schooling as the educational backbone of the enlightened state has – hardly questioned anywhere – survived into post-modernity. And it is only very slowly that the exclusionary authoritative claims of academic interpretations are weakened in school and elsewhere. Still many parts of European societies attribute to academics, to their reasoning and (hi)stories the ability to judge divergent and competing (hi)stories and to make truth about the past. While in post-modernity a narrative transgressing the boundaries of its peer group should be a matter of negotiation and meaningful adaptation, we as academics frequently still withdraw to our established discursive power – and, honestly, mostly we like it as it keeps us in the favourable social position of a priesthood with special access to an arcane, true past.[6]

Returning to the case of Götting the editors of the village chronicle, in a letter to the contributors of March 15th 2007, explained that

> "[t]he celebration of the jubilee [of the 1200th anniversary of the church of Götting being mentioned for the first time] shall strengthen the consciousness of the villagers of Götting for the rich past of their hometown" ("*Die Feier des Jubiläums will das Bewusstsein der Göttinger für die reiche Vergangenheit ihres Heimatortes stärken*").

And here trouble started: Myself being an academic, an archaeologist with some interest in the Middle Ages, grown up in Munich, hardly able to speak the local Bavarian dialect and probably in the Mangfall valley not more than ten years ago for the first time – how could I say anything on the pre- and protohistory of Götting, on "the rich past of their hometown", which has any meaning or importance on the ground? Of course I could tell one or more academic stories, but why should they be printed in the village chronicle? Why only my story? How could I risk to submit the villagers of Götting to my interpretation of their past by superimposing my perspective via a printed narrative in their chronicle? Black-hearted and somewhat pointed, such an act could be described as "academic colonialism" … notwithstanding that "they" had asked me.

---

6 I am most thankful to Mark Pearce, who forwarded this view at our joint Round Table at the EAA-Meeting in The Hague in 2010.

Would it not be more appropriate to a village chronicle contributing and strengthening the local identity of the villagers of Götting that the locals themselves select, which events and places are important to them and should be passed on (cf. HODDER 1999; BEER 2010)? Initially I considered making a poll among the villagers which monuments and traditions are important to them, but soon dismissed this idea due to shortage of time and money. Instead I hoped that local legends might lead my way to those archaeological sites which receive special local importance. Moreover legends are handed on for a long time and thus not so much reflect actual opinions but are an essential part of the shared, collective memory of a place (conceptually on collective memory HALBWACHS 1925; 1950; ASSMANN 1997: esp. 29–160; 1999). I was deeply disappointed to learn that none of my contact persons[7] knew any local legends at all. Nevertheless my approach produced a fully unexpected result: I was offered a colourful miscellany of stories, which do not meet the conventional idea of a legend, but which perfectly met my purpose by pointing to selected places which are of special meaning to the local collective memory. Moreover the analyses of these stories provide much better insights into the modes of interaction and appropriation between academic and local stories than any legend could have done.

I will start my case study with three places in the vicinity of Götting. They show quite different (non-)interactions between academics and locals. Though a case study they will allow for more general questions to be asked: Most prominent in shaping (hi)stories of Götting is one figure – Karl Braßler. He paradigmatically sheds light on the opalescent concept of the "mediator". But while he died in the 1960s his narratives are still highly influential today – leading to considerations on the modes and conditions for establishing successful narratives. Among the modes the authoritative power of school and the established connection of academic writing and truth seem to be of great importance. As a detailed analysis of academic-public-knowledge flows for one of the places demonstrates, these modes of truth-making – even when married together – are far from being successful, if their narratives do not meet further conditions: (hi)stories seem to need places to be attached to and they equally need to meet a broader educational background to be meaningful. Their transformations between archaeologists and

---

[7] My local informants were the editors and the local contributors to the chronicle KLÖCKER – FUCHS (2008), whom I asked at the editorial meeting of November 22nd 2007.

the locals can be understood as appropriations of academic narratives to local frames of meaning, as acts of making meaning.

As Alexander Gramsch has remarked on a draft of this paper, I myself seem to rely more on academic narratives, being more interested in their public appropriations than in a feedback of public narratives into academia. On the one hand he is right! Being an academic I am wearing these glasses and looking on the world through them. In each of the case studies I will first deliver the archaeological perspective, followed by an analysis of the local (hi)stories. However, writing an academic paper, I am not giving a plain account of local perspectives, but I am instantly going to analyse them according to historical plausibilities and academic rationality. In a way I am appropriating these local (hi)stories by making them objects of torture by my analytical tools. On the other hand I would have loved to find any such feedbacks, but – except for two small notes[8] – no concrete feedback came to my attention. On a more general level, however, this paper itself may of course be regarded as a feedback as has been the village chronicle of 2008 with its initiative by the villagers.

Academic and local (hi)stories

No interaction at a "Gothic" female grave

In the archaeological world of the Migration Period Götting is quite well known for a female burial, accidentally found by the villager Josef Rindele in the southwestern corner of the borough gravelpit in 1940. While removing grass sods he came across a skeleton, accompanied by a pot, a silver fibula (a second one was probably overlooked) and a bone comb (*Fig. 1*). The local historian Karl Braßler first published the finds in the local newspaper two years later (BRASSLER 1942) and again in a local journal and a newspaper in the late 1950s (BRASSLER 1957: 12f. and opposite p. 17; BRASSLER 1959b). These publications, however, did not help much for the popularity of the finds as the news neither impressed locally nor were they received in the scientific community. The academic interest did not change until a new publication by archaeologist Erwin Keller, later on member of

[8] Karl BRASSLER (1957: 8), while dealing with the barrows of Unterleiten, not only refers to archaeological finds but also to local knowledge as it is stored in the old field name of the site (cf. *infra*).
STEFFAN – UENZE (2003: 138f. no. 153) shortly consider a possible Roman past of the Stiegelburg, which so far had only existed in local tradition (cf. *infra*).

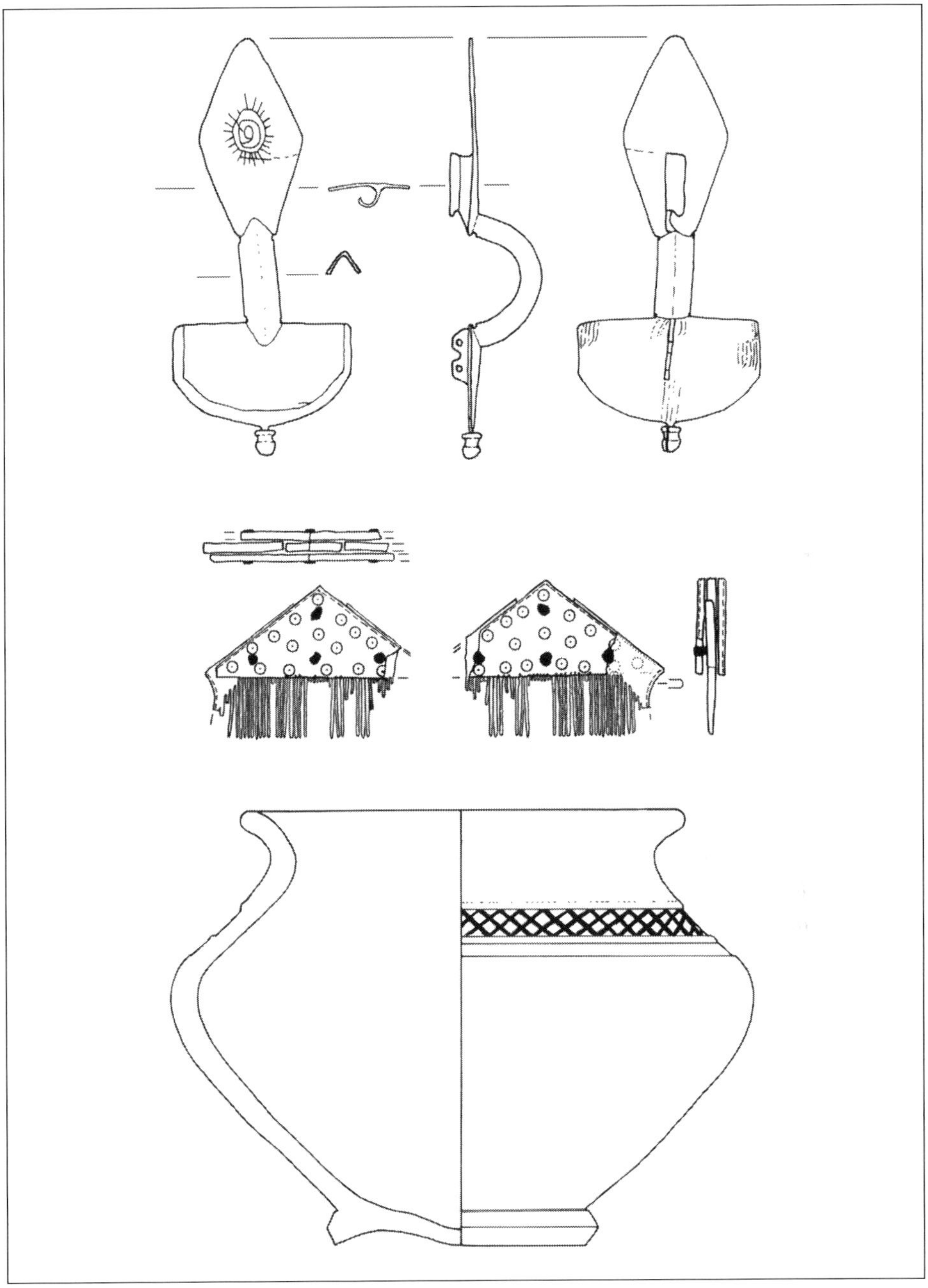

Fig. 1. Pot, fibula and comb from a female burial found close to Götting in 1940 (© Thomas Meier after KELLER ((1971) with modifications).

the Bavarian heritage board, in a regional archaeological journal (KELLER 1971). In 1988 the finds were on display at an exhibition on the Baiuvarians, officially promoted by the states of Bavaria and Upper Austria (DANNHEIMER – DOPSCH 1988: 372 no. M I.7). Obviously it was here that the actual homeland researcher Helmut Loose saw and 10 years later integrated them into his chronicle of Götting (LOOSE 1998: 7) – with some mistakes[9] and again a total lack of reception on behalf of the villagers. Even the integration of the finds in the local museum does not help this neglect of local attention.

According to the scientific convictions of his time Keller concentrated on the fibula. Being part of the traditional female costume its form seemed to be significant for specific ethnic groups, i.e. Germanic tribes known from written sources. Under this paradigm the Götting fibula perfectly matches the form attributed to the Goths. As antique authors convey that the Visigoths moved from the Crimea via northern Italy to Aquitania during the first two decades of the 5th century (in summary BIERBRAUER 1994), the Götting grave seemed to prove that not all Goths migrated via Italy but some groups passed the Alps on their northern rim (cf. KELLER 1987: 77f.), while the 1988-exhibition connected it to late Roman military presence in the region as auxiliary forces often integrated Germanic warriors (BÖHME 1988: 26–29).

When re-evaluating the grave I have tried to show that it can be dated quite precisely to the years around 400 AD (MEIER 2004b), but an ethnic interpretation in the sense of migrating peoples today seems out of date as it is obviously guided by the national-folkish ideas of the late 19th century and is only applicable to the age of nationalism (cf. BRATHER 2000; but BIERBRAUER 2004). I do not want to delve into the depths and shelves of ethnic interpretation in archaeology, but remarkably enough those people who buried the woman close to today's village of Götting did so in quite an unusual fashion: seemingly they regarded her as a stranger, not fitting into the local identity and requiring other habits than usual, and in some way associated her – correctly or not – with those Balkan tribes called "Goths" (for a detailed discussion cf. MEIER 2004b).

What makes Götting quite famous in the archaeological world, however, until today did not make its way into the common knowledge of the villagers – despite its publication and reception by local historians Braßler and Loose.

---

9 The fibula is said to be of bronze (instead of silver) and the comb he thinks to be of wood (instead of bone).

Probably the woman from the gravel pit was not alone as a wider "Gothic" presence may be reflected in the place name of Götting. The name is recorded for the first time as *Cotingas* in 776/83 (BITTERAUF 1905: 104f. no. 83; WALLNER 1950: 78). Usually it is construed from a mythical founder of the village called **Coto*, a personal name otherwise completely unknown. Would it not be equally or even more plausible to derive the place name from the "Goths" (cf. GOTTHARD 1849: 6f. [but not mentioning Götting]; FINSTERWALDER 1952: 41)? In this case the identity – self-concept and/or external perception – of the woman from the 1940 grave and her kinsmen would be at the beginning of the local identity of today's Götting and would have been written into the collective memory of the place by its very name. It is this Gothic interpretation of the place name, which Helmut Loose adopted in his popular chronicle of Götting (LOOSE 1998: 7f.) – but as the woman's grave remained without consequences on a local level: Reading the name of their place the villagers widely follow the *Coto*-hypothesis instead of an allegedly "Gothic ancestress".

Lost interaction at the Hallstatt barrows of Unterleiten

Only in oblique light and favourable soil conditions do the shallow remains of some barrows east of the hamlet of Unterleiten, which is an administrative part of Götting, become visible from the air (*Fig. 2*). According to an air photograph originally there were at least twelve barrows of which only nine were still visible in 1886 (WEBER 1886: 7f.) and four in 1921 (REINECKE 1924: 94), while by 1957/59 all of them had disappeared (BRASSLER 1957: 8; 1959a: 46f. no. 15). At least for eight of the barrows there is some information from old excavations: the finds are few and not very outstanding dating to Hallstatt D1 (Kossack 1959: 28, 191f. no. 153 pl. 112; STEFFAN – UENZE 2003: 135–138 no. 150 pl. 50.5–8, 52–53, 55.1–2). Groups of barrows like those at Unterleiten were quite common in southern Germany during the Hallstatt period and some of them are very well preserved until today and/or excellently excavated. From an archaeological point of view the barrows of Unterleiten are by no means exceptional.

This archaeological evaluation seems to correspond to a lack of local meaning of the site. In the course of my research, however, I have come across a "real" legend connected to these barrows: In the late 1950s, i.e. approximately a generation after the last barrows had disappeared, Karl Braßler still knew the old field name "*Im Schatz*" ("In the treasure") (BRASSLER 1957: 8; 1959a:

Fig. 2. Air photograph of a small group of Hallstatt-barrows east of the hamlet of Unterleiten (© Bayerisches Landesamt für Denkmalpflege-Luftbilddokumentation, Aufnahmedatum 01.12.89, Foto Otto Braasch, Archiv-Nr. 8136/033, Dia 6028-19).

46f. no. 15).[10] In Unterleiten he was told that at the place where the barrows once stood there is still buried a golden chest in which a sultan was interred (BRASSLER 1959a: 49). Field name and legend most probably correspond to each other and we might reasonably assume that both are based on prehistoric metal objects, which are regularly found in such barrows. As most of them are made from copper alloys it is a usual confusion to take them for gold and thus exorbitantly exaggerate the "treasures" to be found in burial mounds. Maybe it was the common knowledge of "treasures" and bones in barrows, which was at the core of the field name and legend at Unterleiten, maybe it was a reflection of the early archaeological research in some of the Unterleiten mounds. In any way the treasure-motif appropriates and transforms a typical archaeological source into a local story giving meaning to a site and making it a place (TUAN 1977). Remarkably enough a homeland researcher like Braßler, interested in almost everything in his small world of the lower Mangfall valley, draws on the legend and the field name to illustrate an old knowledge about the burial mounds and

[10] Today this field name is lost and the area is called "Rosssteig" ("Steed track").

the grave goods within, thus mirroring a public appropriation back into academic interest.

It remains unexplained how the sultan found his way into the barrows of Unterleiten. Two ways come to my mind – the second with kind help of my colleagues and a glass of wine: Perhaps the sultan reflects the fears with the Great Turkish war and second siege of Vienna in 1683. These events happened far to the southeast of Bavaria, but the Bavarian Elector Max II Emanuel and his troops took part in the Imperial army relieving Vienna in October 1683. After their victorious return the soldiers might have spread rumours of the formidable treasures of the Turks making the "sultan" a kind of synonym for the owner of a really great treasure – and thus burying him in a barrow at Unterleiten. Or the sultan is simply a synonym for a heathen as burial mounds usually were regarded to be of pre-Christian date (which most of them are from an archaeological point of view as well).

Transformative interaction at the Stiegelburg

The Stiegelburg hillfort is situated on a spur of the Irschenberg southwest of Unterstaudhausen, another hamlet in the vicinity of Götting (*Fig. 3*). The core of the site is quite small, measuring no more than 20 m in length. It is slightly elevated against the outer keep in the south and west towards which it is demarcated by a wall and a ditch. The outer wall and ditch are today mostly levelled by agriculture and forestry. Written sources regarding the Stiegelburg are missing, as are archaeological excavations.[11] Our knowledge is merely based on some strayfinds of Middle Bronze Age and Urnfield pottery (16th–11th century BC) and some medieval sherds (12th–14th century AD) (Archive of the Bavarian Heritage Board). All of the prehistoric pottery is secondarily burnt and accompanied by heavily burnt and highly fragmented animal bones (Weber 1909: 167; Brassler 1957: 9f; 1959a: 56 no. 31, 69, 77 bottom right, 80 no. 36; Steffan – Uenze 2003: 138f. no. 153 pl. 41.6, 8–10).

Archaeologically the Stiegelburg is one of some hundred pre- and protohistoric hillforts in Bavaria (cf. Leidorf – Ettel 1999). The finds indicate that the site – as in many other cases – was reused and restructured in different periods. The

[11] With some academic arrogance I do not consider the excavations to be archaeological, which Karl Braßler planned for in a newspaper article in 1943 and which were obviously conducted some time before his next article in 1950. Further information about these digs is missing.

Fig. 3. The area of the Stiegelburg at the foot of the Irschenberg southwest of Unterstaudhausen (© Bayerisches Landesamt für Denkmalpflege-Luftbilddokumentation, Aufnahmedatum 14.04.86, Foto Otto Braasch, Archiv-Nr. 8136/068, Dia 4344-25).

burnt pottery and animal bones suggest that once in prehistory the Stiegelburg might have been a place for burning sacrifices (*Brandopferplatz*) which is a type of sanctuary quite well known in Southern Germany and the Alps from the Bronze Age to Roman times (WEISS 1997; GLEIRSCHER 2002). Like the barrows of Unterleiten the Stiegelburg is quite ordinary from an academic point of view – at least on the basis of the actual almost non-existent state of research.

This unflattering academic evaluation is opposed by the local meaning of the Stiegelburg. Among the villagers of Götting it is the most prominent archaeological site to be mentioned, it enjoys great interest and is well anchored in a detailed collective memory. According to my local informants[12] the place name "*Burg*" indicates a (medieval) castle, which corresponds to the walls and ditches on site. But long before the castle there was a Celtic sanctuary; its age is indicated by

[12] Cf. *supra* footnote 7.

the adjacent field structures of the ridge and furrow type. Roman coins hint at a watchtower. Finally the stones of the medieval castle were used to build either the church of Westerham or the castle of Pullach, both in neighbouring villages (cf. GIESE 2006: 14).

The graphical counterpart of this knowledge is a mural by Antje Mayer in the assembly hall of the school in Götting made in 1969 (*Fig. 4*). It is titled "Celtoroman fortification around the birth of Christ" ("*Keltoromanische Befestigung um die Zeitenwende*") and shows a *murus gallicus* as a typical Celtic wall with warriors in typical Celtic clothes. On the left hand side the wall is already finished, on the right hand side it is still under construction, which, on first sight, gives the impression of a tower. In the background there are amongst other things, a megalithic tomb and long, small fields, probably ridge and furrow. In front of the wall there are some prehistoric pins, a fibula, a knife and an axe, all of them dating to the Bronze Age.[13] The fortification is unnamed, but quite obviously it is intended to be the Stiegelburg.

Among the stories of the Stiegelburg the transfer of its stones to Westerham or Pullach appears quite legendary, not the least as this motif is rather common. Interesting enough it is this transfer of stones which we can trace back furthest: While the Westerham-version was recorded a century ago according to oral tradition of an old man (GRAF 1907: 24, 55), the Pullach-version allegedly was excerpted from old files in the first half of the 19th century (GRASSINGER 1857: 177). The "factual" credibility of these traditions is very doubtful, but what is much more interesting is the ideological connotation that a transport of stones is not only a transfer of building material: together with the stones, authority and legitimation are transferred as the material builds a bridge back to the quality and lordly rights of its old place. At the time when the stones of the Stiegelburg came to Westerham or Pullach – or when these traditions were constructed – the Stiegelburg must have been highly esteemed so that it was a benefit to connect new sites to this old place. Interestingly enough the factuality of the legendary transfer of stones is discussed (with different enthusiasm) in several studies of local history and connected to long forgotten rights of possession (GRASSINGER 1857: 177; GRAF 1907: 24; BRASSLER e.g. 1956).

The other stories of the Stiegelburg are much easier to trace back: the prehistoric sanctuary is based on the fragments of pottery and animal bones mentioned above.

---

13 Based on BRASSLER (1957: upper plate opposite p. 17; 1959a: plates on p. 55, 59 and 77).

Fig. 4. Mural by Antje Mayer (1969) in the assembly hall of the school in Götting, titled "Keltoromanische Befestigung um die Zeitenwende" *("Celtoroman fortification around the birth of Christ") (© Nicolas Klöcker).*

Finds of the Celtic period, however, are missing[14] but Karl Braßler argues that the neighbouring fields with ridge and furrow are Celtic and thus proof a contemporary use of the place of the Stiegelburg (BRASSLER 1956; 1959a: 75–78, 80 no. 36; again LOOSE 1998: 6). Today – after long academic discussions – the ridge and furrow fields of Southern Germany are regarded as typically medieval (SCHWARZ 1989: 199ff. esp. 199–201, 221; but FRIES 1995: 126ff. for examples from the Iron Age onwards) while neighbouring field structures are anyway inappropriate to date the walls and ditches of the Stiegelburg as the fields might be oriented along older structures. Both archaeological arguments against a Celtic sanctuary, however, did not find their way into the local knowledge.

Like the Celts the Romans, too, are Karl Braßler's product: He argues this on the basis of some fragments of pottery, which he thought to be *terra sigillata*, and some Roman coins, which are missing today (BRASSLER 1956; 1957: 10). While the pottery is now dated to the Middle Ages (Archive of the Archäologische Staatssammlung München, handwritten correction to inventory no. 1952, 793) Roman coins – taken that they were really Roman – are found in almost any context in any place and therefore without further evidence cannot prove a Roman settlement. The watchtower was cautiously considered by Karl Braßler in 1941 (BRASSLER 1941), but he was obviously more convinced of it two years later (BRASSLER 1943) while after World War II he only mentions the strategically important position of the Stiegelburg on a spur in a more general way (e.g. BRASSLER 1956).

A for now final step is the globalisation of Braßler's (hi)stories via the internet as they became part of the greatest project of shared knowledge humankind has organised so far: in the German Wikipedia there is an entry on the Stiegelburg, (re-)written by at least twenty authors, stating that the castle was probably erected by the Celts, but was surely used by the Romans and the Bajuvarians as is proved by four copper-coins, a spearhead and some fragments of pottery found during excavations. According to these opinions the medieval castle served to control a road, but it is unknown when the castle was destroyed. Some local chronicles report that the stones were used in the castle of Pullach and the small church of Westerham. The age of ridge and furrow-fields in the vicinity may be medieval or much older (http://de.wikipedia.org/wiki/Stiegelburg [31.12.2012]). From Wikipedia this knowledge spreads into further homepages

[14] BRASSLER (1956) mentions an iron spearhead and an iron spatula from Celtic times, but soon – and *archaeologically* correctly – remarks that they could be medieval as well (BRASSLER 1959a: 69).

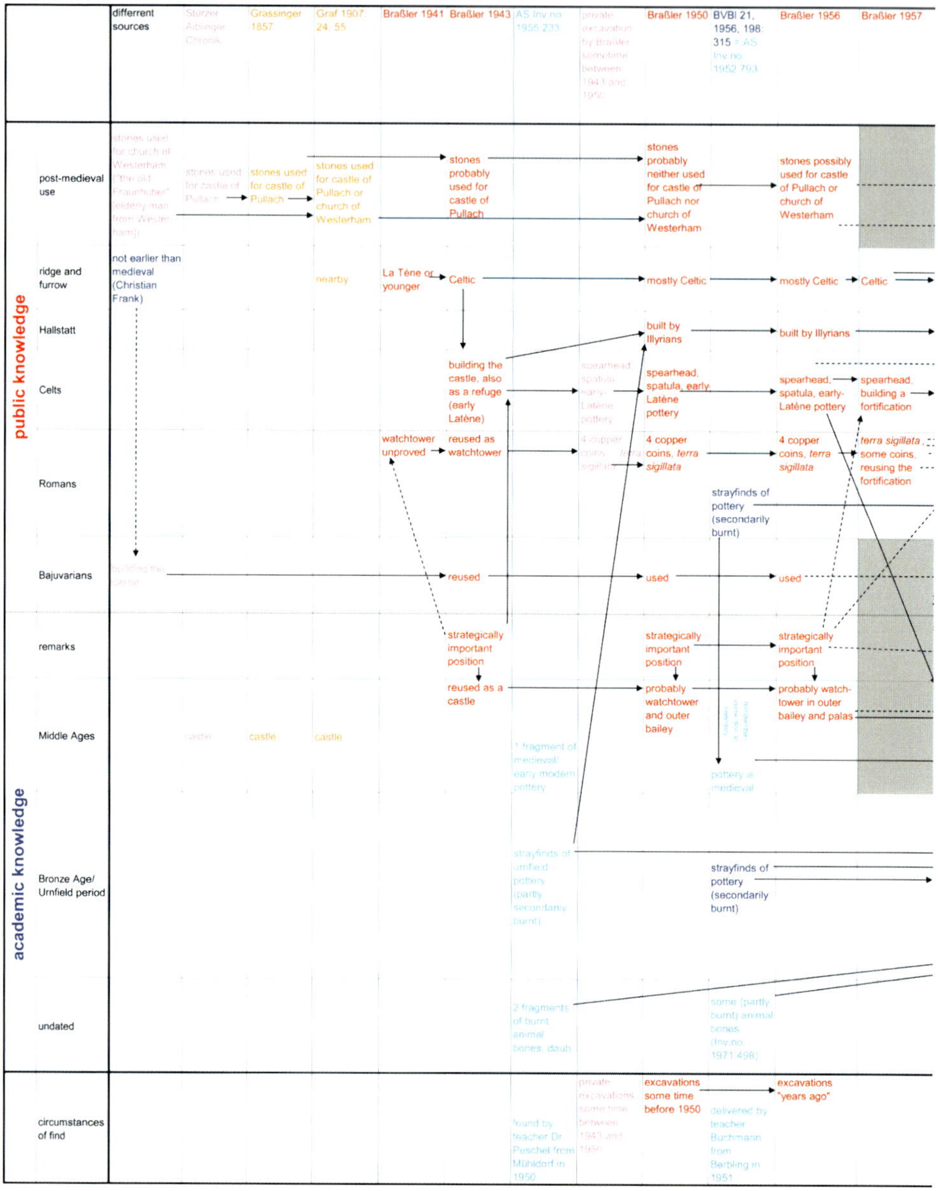

(e.g. http://www.bnro.de/~dfuchs/umweltgruppe/Exkursionen/berbling.html [24.3.2008]). Though this information is still clearly based on Braßler's works there are some slight changes: obviously slowly some doubts arise about the Celtic age of the Stiegelburg (and the fields nearby) while the Bronze Age items and the sanctuary got lost (as they are already missing in the school-version).

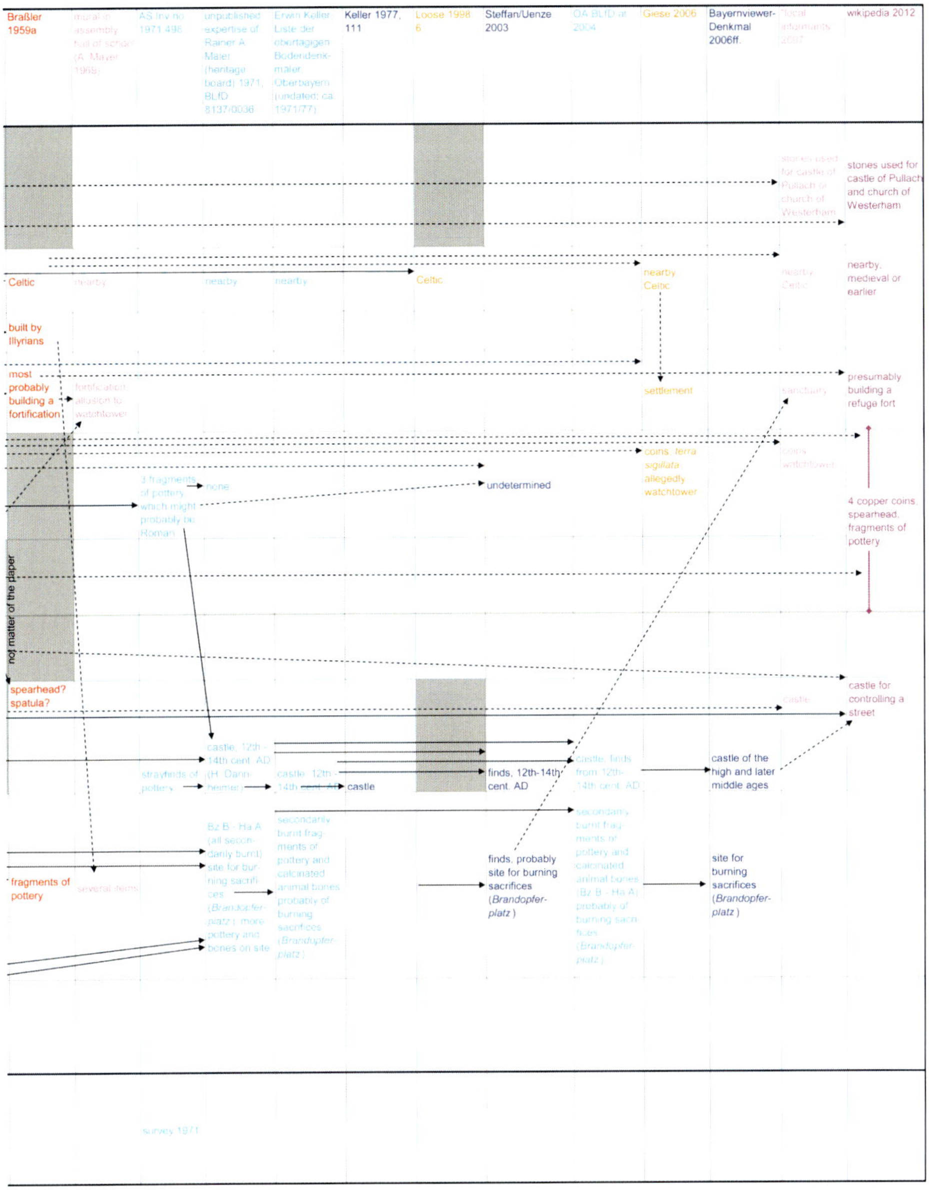

Fig. 5. Flow of knowledge on the Stiegelburg (deep blue: academic publications; light blue: academic notes in archives and files; deep red: publications by Karl Braßler; orange: local publications; light pink: unpublished local material/actions; deep pink: wikipedia; grey shading: not the matter of the paper; arrows: attested flows; broken arrows: assumed flows).

In short we may simply call the Stiegelburg-story a selection and transformation of (para)academic knowledge. But this is not very illuminating if we want to delve into the depths of academic-public-interactions. *Fig. 5* sums up the flows of knowledge around the Stiegelburg: At first glance the chart looks quite chaotic, but on closer inspection the two discourses of local knowledge (red shades) and academic knowledge (blue shades) are surprisingly clearly separated, i.e. the interaction between the two spheres is not as close as assumed.

Selective knowledge (un)flows

Local knowledge of the Stiegelburg is characterised by some long and stable traditions, i.e. a Celtic origin of the site due to the Celtic date of the nearby ridge and furrow fields, the Roman watchtower and the Baiuvarian/early medieval use, although the last point was never fleshed out in any detail. Neither of the three points is valid (any longer) in academic (hi)stories, but such disapproval is without influence. This, however, does not imply a total neglect of academic reasoning, as Braßler used a (private and illegal) excavation and thus a scientific, positivistic method to prove his assumptions of a Celtic origin of the Stiegelburg. Indeed he found objects, which he thought to be Celtic (and Roman) and thus had material witnesses in his hand. Clearly these finds demonstrate that facts are not objective but made by anticipation. Interestingly enough he does not say anything about medieval finds from his excavations, although they would be expected in high numbers, but obviously were out of his interest (and sight?). On the other hand it might be a telling coincidence that Braßler's finds are not only lost, but in the reception of his Celtic narrative the finds do not play any significant role. This also applies to the Bronze Age pottery from the site, which Braßler mentioned in his last paper (BRASSLER 1959a) but without traceable impact on the villagers. Even for the medieval castle, which is the undisputed common ground of local and academic knowledge, public lore is based on the – quite well preserved – material evidence on the ground, while the – admittedly unspectacular – finds do not play any significant role.

While finds – regardless of the unimpressive sherds from the Stiegelburg or the quite spectacular fibula from the "Gothic" burial – generally do not play any role in the common knowledge of the village, remarkably enough knowledge of the Roman finds has been passed on. This Roman narrative is even so suggestive that again and again it has entered the academic perception of the site: Not only were some of the private stray finds from the early 1950s regarded and published as Roman and later on in the 1970s some of the sherds again were classed as possibly Roman – probably under the convincing influence of the finder and little knowledge of medieval pottery at that time – but even after the expertise that none of the finds are Roman (from an archaeological perspective) the latest academic publication (STEFFAN – UENZE 2003: 138f. no. 153) leaves the door ajar for a Roman past of the site.

In the academic sphere especially the Bronze Age- and Urnfield-narrative developed step by step from the determination of the pottery via the secondarily burnt status of the sherds and the animal bones to a prehistoric site for burning sacrifices (*Brandopferplatz*). This interpretation was forwarded in 1971 for the first time, but it took until 2003 before it was published. From this publication or the web-based Bayernviewer-Denkmal (www.blfd.bayern.de/denkmalerfassung/denkmalliste/bayernviewer/index.php [27.2.2013]) the sacrificial site was immediately received in Götting, as my informants already knew about it in late 2007. However, they did not receive its dating to the Bronze Age or Urnfield culture, but attributed it to the Celts. The Bronze Age, EU-commonly propagated as the first Golden Age of Europe (cf. NIKLASSON in this volume: 63), obviously misses its audience ...

Unflatteringly the chart also demonstrates that the academic formation of knowledge to a great part is happening internally for example in local files (*Ortsakten*) or inventories and often remains unpublished for decades (but this may change with the entry of the digital age into heritage management). On the one hand these different practices of publication, with the bulk of small and highly redundant articles in local newspapers and journals versus the silence of academic archives mirrors (or produces?) the different importance the Stiegelburg is attributed by the locals and by academics. On the other hand Braßler (and others) had and used the opportunity to get access to the information of the heritage board and the museums and thus to draw on unpublished academic knowledge, but did so only once in the early 1950s with the Bronze Age stray finds friends(?) of Braßler had delivered in Munich. Braßler adopted the dating of the pottery to the Middle Bronze Age to Early Urnfield culture, and according to the phenotypical classification of races (*Rassenkunde*) and ethnic-historical paradigms of the 1920s and 1930s concluded that it must have been the so called Illyrians, who first built a hillfort on the site of the Stiegelburg. In Götting[15] none of those "Illyrians" was ever acknowledged outside Braßler's writings ...

---

[15] Generally the Illyrians were highly *en vogue* during the 1920/30s (e.g. SEGER 1926). I am especially thankful to Karin Reichenbach for this hint!

The three examples discussed so far are quite different in matters of academic-public-interactions. They differ in matters of interest, knowledge and intensity of appropriation. They raise many questions on the conditions of appropriation and mechanisms of transformation, which I will discuss on a more general level in the second part of this paper. But – beside being situated in the vicinity of Götting – the common ground of all three examples is the central position Karl Braßler takes in current (hi)stories on Götting's pre- and protohistory. At first I will devote to him an analysis as a "typical" mediator. Or not?

Karl Braßler – Academic? Mediator? Public?

Living himself in Götting for a long time Karl Braßler is a very chatoyant and highly problematic figure: born in 1897 he was for a short time student at the University of Munich,[16] but mainly worked as the leader of the student group of the *Deutschvölkischer Schutz- und Trutz-Bund* and the most active anti-Semite agitator and NS-propagator since 1920 (LOHALM 1970: 168). After he was arrested he returned two years later as one of the early members of the NSDAP (LOHALM 1970: 169) and took part in the *Hitler-Putsch* of 1923; later on he was member of the *Blut-Orden* honouring those who already followed Hitler in his early days. In the 1930s he led a NSDAP location group in Berlin, but moved to Götting in 1939 and for some dubious reason was dismissed by the NSDAP in 1942/43. From 1943 to 1945 he worked as a parish clerk in Götting as obviously with his dismissal from the NSDAP he did not fully fall into the party's disgrace. After World War II we find him as a private scholar and homeland researcher primarily in natural science and local history (STAHLER 1956: 11; KLÖCKER 2008: 151f.). His engagement and vision in those years was honoured by the Cross of Merit on ribbon of the Federal Republic of Germany (STAHLER 1956: 12). Although Braßler's interest was not primarily or not even to a high degree in the archaeology of the Mangfall valley, he published a tremendous lot of small notes and papers in his Götting years from 1939 to his death sometime in the 1960s.

16 Braßler was enrolled at the Ludwig-Maximilians-Universität of Munich from fall term 1917 to summer term 1920 but in 1918 he was on military service for two terms and the following two terms he was suspended; there only remain fall term 1917 and summer term 1920 for an active attendance. He studied at the faculty of philosophy, which at that time comprised everything from philology, history and classical archaeology to physics, geology and botany (Personalstandsverzeichnisse der Ludwig-Maximilians-Universität München at epub.ub.uni-muenchen.de/view/lmu/pverz.html [27.2.2013]).

His publications are not only again and again covering almost all monuments in the vicinity, but Braßler managed successfully to place his writings in daily newspapers and local journals so that he developed an area-wide and permanent presence of his ideas and interpretations.[17]

While Braßler's role in shaping the historic traditions of Götting is quite clear it is hard to typify him. On the one hand he surely is a mediator, passing on academic knowledge – though often enough of problematic political background – to the wider public. With his predilection for publishing in local newspapers and journals he strongly emphasized his efforts to reach, to educate and to convince the public of his version of their history. It is mainly this role in which I considered him in the three examples above.

Braßler, however, did not receive his (hi)stories from "high" academic publications, but mostly constructed them himself, in many cases doing genuine research. With an academic background he obviously had some basic archaeological and historical knowledge at the level of the 1920s, which he did not update in the later years of his life.[18] The use of an excavation to prove his hypotheses on the Stiegelburg demonstrates his academic approach, as does his historical discussion on a possible reuse of the stones from the Stiegelburg at the castle of Pullach and/or the church of Westerham despite of (or due to?) the legendary structure of this tradition.

In a simple model of "academics – mediators – public" Braßler is a mediator in his own right and simultaneously takes the academic part of producing the (hi)story. Finally, regarding the greater contexts of his narratives he drew on bits and pieces of academic knowledge, appropriating them like any other member of the public for his own purposes as e.g. he took the supposedly Celtic date of the ridge and furrow fields for dating the adjacent Stiegelburg. As Michael Cremo points out for his own person (CREMO this volume) – but surely for very different agendas! – the categories of "academic", "mediator" and "public" are melting down in the person of Braßler as well. It is only a matter of the analytical perspective as to which category he is assigned.

---

[17] (E.g. in the paper BRASSLER 1957) he lists 64 titles from his hand and this was not his last publication.

[18] It is surely too simple if KELLER (1971: 169 note 4) classifies Braßler as a layman in matters of archaeology in order to rule out/ignore Braßler's observations and interpretations of the "Gothic" burial.

Beyond problems of typifying a complex individual like Braßler, a more interesting aspect are the truth values assigned to his narratives on behalf of archaeologists and his fellow villagers. While archaeological receptions to his work stick to simple references to sites and finds mentioned or published for the first time in his many papers, they completely ignore the greater narratives in which Braßler had wrapped the archaeological "data" (e.g. KELLER 1971: 169; STEFFAN – UENZE 2003: 135, Nr. 150 [barrows of Unterleiten]). But it is just these narratives which are widely accepted and highly esteemed in Götting – even today. Exemplarily Helmut LOOSE (1998: 4) in the preface to his chronicle of Götting argues that he will not go into depth with the prehistory and medieval time of the village as its history up to the Thirty Years War has been extensively covered by Braßler in 1957, who "has collected with greatest diligence and meticulousness for the history of Götting" ("*der mit größtem Fleiß und Akribie für die Geschichte Göttings gesammelt hat*").

The vivid validity of Karl Braßler's version of Götting's history even half a century after his death is even more striking as this vision of the past is rooted in the very right-wing nationalistic world-view of the 1920s. It is surely not enough to explain such longevity by the lack of a more recent homeland researcher, who engaged with Götting's past with the same enthusiasm. The permanent presence of Braßler's visions of the past in local newspapers was surely an important device to set them up in the collective memory of the villagers. But Braßler was not the only one with such presence in the media and – analytically even more important – not all of his stories were received with the same enthusiasm (to say the least). Conversely we have to consider, why until recently none of the academic counter-arguments to Braßler's version made its way into local knowledge. Which historical (hi)stories or which parts of them are appropriated locally obviously not only depends on persons and media, but we have to look for further factors.

Making truth

If narratives – as I have stated in the introduction – are constructed according to rules and plausibilities of individual social groups and primarily are bearing meaning within the horizon of their groups, it calls for explanations why some narratives (or parts thereof) are more successful than others in crossing the borders between social groups, and why some gain meaning far beyond their group of origin.

Fig. 6. Sgraffito by Antje Mayer (1969) on the southern façade of Götting primary school (© Nicolas Klöcker).

The Götting examples hint to modes of truth-making as one important factor for the inter-group success of narratives.

Götting primary school (1969): The making of ordered history

The Stiegelburg-mural in the assembly hall of Götting primary school (*Fig. 4*) was painted in 1969, some years after Braßler's death. It is accompanied by a second mural on the corridor of the first floor (today the school's computer-room) depicting the Rosenheim lake with some fanciful pile-dwelling village.[19] A third sgraffito on the southern façade of the school (*Fig. 6*) depicts four scenes from the middle ages and early modern times probably starting on top with the setting up of the village in a forest by the famous **Coto,* continuing through a jousting knight – probably recalling the counts of Falkenstein as a regional noble family of the 12th century – a ransacking band of soldiers – probably from the Thirty Years War – and ending at the bottom with some bucolically grazing sheep and cows (today?).

These pictures document the local dominance of Braßler's stories, which they monumentally convert into images. Simultaneously they highlight the reception of Braßler by his fellow villagers: given that the school context of the pictures did not allow for fantastic visions, but required an academic approval of the motifs and their details, the murals illustrate the local esteem of Braßler as an honest researcher who truthfully mediates academic, reliable knowledge.

To evaluate the full impact of the murals, however, we have to consider the school context of the paintings more closely. As attending school is an obligation and school has (almost) a monopoly on the transfer of "true" knowledge and education, narratives which have successfully invaded school curricula, are armed with legal coercives (school marks), potentially decisive on the further course of life. These narratives are equipped with the highest social assertiveness

[19] The Rosenheim lake was fed by melting waters of the Inn-glacier at the end of the last glacial period. It covered the lower Mangfall valley approximately to the western end of Götting and was mostly sedimented and partly dried out ca. 10.000 BC. Karl Braßler according to the geographical knowledge of his days was convinced that the lake still existed in neolithic times and silted up during the Bronze Age, so that the first settlers of the region could build their houses only on neighbouring hills or as pile-dwelling villages at the shore of the lake – probably on the spot of later Götting where there was found a copper vessel allegedly from the Copper Age (e.g. BRASSLER 1950; 1957: 7; 1959a: 33f.).

and may aptly be called "ordered history". Thus school is – generally speaking – an institution to assert discursive powers (FOUCAULT 1975), to standardise behaviour and to re-produce social habitus (BOURDIEU 1967: 138f.; 1979). The position of the murals in focal points of the school building – confronting every child with Braßler's vision of the past every day – tightly binds them to the school as an authoritative institution of knowledge transfer. Though I have not attended lessons taught on local history in the school of Götting, it might be telling that Helmut Giese not only was a teacher there for almost 30 years (1976-2005) (GIESE 2008: 292), but also takes care of those parts of Braßler's archive, which are kept in the museum of nearby Bruckmühl. His short historical summary in a recent publication on the borough of Bruckmühl follows Braßler's vision of the past in every respect (GIESE 2006: 10, 14). Therefore, we might assume that Giese has taught Braßler's (hi)stories in school as well – the spoken word of the lessons and the murals mutually illustrating and explaining each other. Moreover, the materiality of the depictions on the walls emphasises their lasting validity as being literally set in stone; their permanence can be sensually experienced and their immutability becomes manifest. On this background it is little surprise that Karl Braßler's visions dominate the perception of the past in Götting until today.

Calling for academia (2007): The revelation of truth

As far as written sources have survived the church of Götting was mentioned for the first time in 809, when it received some slaves and properties (BITTERAUF 1905: 253 no. 292 [wrongly attributed to the village of Gauting]; SEPP 2008: 191–194). When the 1200th anniversary of this donation approached the parish church council took the initiative to celebrate the jubilee and on that occasion aimed to produce another chronicle of the village. Soon it was decided that this volume should not be a reprint or summary of Braßler's visions of the past, as a very abridged version had been published shortly before (LOOSE 1998: 6-9; GIESE 2006). Instead the matter should be approached with "academic accuracy" (KLÖCKER – FUCHS 2008: 12). Two of the villagers became the editors: Alois Fuchs, member and long-time chair of the local parish church council, and Nicolas Klöcker, at that time student of history and German philology at Munich University. The latter, well-connected with some younger researchers in regional history, assembled a crew of academics (including myself) and academically trained locals to contribute to the volume. In a letter to the contributors of March 15th 2007 the editors defined the aims of the miscellany that it

> "shall be academically founded and simultaneously comprehensible for a multilayered readership. [...] The celebration of the jubilee shall strengthen the consciousness of the villagers of Götting for the rich past of their hometown. You fundamentally contribute to this by your paper as you make historical connections accessible to a broad public" ("*Der Sammelband soll wissenschaftlich fundiert und zugleich für eine vielschichtige Leserschaft verständlich sein.* [...] *Die Feier des Jubiläums will das Bewusstsein der Göttinger für die reiche Vergangenheit ihres Heimatortes stärken. Einen grundlegenden Beitrag leisten dabei Sie mit Ihrem Aufsatz, indem Sie historische Zusammenhänge einer breiten Öffentlichkeit zugänglich machen*").

Obviously the editors and the parish church council were well aware that history is an important means of (local) identity. It remains open as to what had fostered the parish church council's wish to put such a measure of identity formation in the hands of academics. Maybe it was Nicolas Klöcker who convinced them on basis of his academic network, maybe there was a silent desire for academic approval of the established Braßlerian (hi)stories, maybe there were finally some doubts raised about Braßler's vision of the past and even if just on the basis of their age. Implicitly, however, the parish church council adhered to a modern, enlightenment vision of a historical truth, which an academically grounded work will reveal and educate the public to understand the historical connections.

For whatever reason – in an editorial meeting I got the impression that at least some, but far from all attendants felt few doubts about Braßler's (hi)stories – academic support was wished for, the wish itself obviously was based on an ascription of sovereign interpretative authority to academia, assuming they have the power of making historical truth – and thus indeed investing them with this power. But while the ordered knowledge of school is equipped with coercives the efficacy of academic stories is based on more subtle means: they pretend to be grounded in objective, unquestionable facts, often inaccessible to laypersons, as it requires special skills and techniques to make these facts "speak" – camouflaging that it is these skills and techniques in the first place, which situationally and discursively construe those "facts" (SCHAPP 1953; LATOUR – WOOLGAR 1979; for archaeology cf. KARL 2010). These facts are based on and ordered in narratives according to the grand stories of our time (LYOTARD 1979) and thus gain special plausibility. Formally, academic narratives are ennobled by the magic of imprinted writtenness, often in a complicated language and fortified with footnotes. And not the least they profit from the still high reputation of academia in our societies ...

Local rejection and subversive laughter

I was deeply afraid that my academic version of the past – simply because it is academic – might be regarded as the correct, "true" story of the village's prehistory and in a worst case might end up as another mural in the school building! Fortunately enough, about one and a half years after the chronicle of Götting was published in 2008, I received a letter by a person who was interested in the history of his family and draws some connections to Götting. In a lengthy note he responded to my printed paper and frankly admitted: "My first impression: Here again a studied historian tries to tear up an assiduous layman (Mr Brasler)." ("*Mein erster Eindruck: da versucht wieder ein studierter Historiker einen fleißigen Leien [sic!] (Herrn Brasler) in Stücke zu reißen*"). Fortunately enough at a second reading his first impression changed – though he did not say in which direction ...

Apart from rejoicing that my meta-story had worked – there is nothing out there like an authoritative true past – I am deeply thankful to the writer as he has proved that academic (hi)stories are not sacrosanct *per se*, but they may be openly rejected. And they may be rejected simply because they deconstruct (hi)stories by non-academics, the reason of rejection not being founded in differing truth claims, but in questioning established authorities and social hierarchies. Even more, academics may be perceived as arrogant, because they do not take laymen's (hi)stories seriously, despite the fact that these (hi)stories might be much more relevant and meaningful outside academia.

While history might be ordered and pupils might be forced to learn it, probably with the argument that it is academically founded, none of these means and arguments necessarily creates meaning for those subordinated to the authoritative view of the past. It is a sign of post-modern plurality that such resistance against alleged authorities can be expressed openly, but the multiplicity of stories is far from new. In more authoritarian contexts, however, resistance against ordered histories and questioning of authorities in charge sought other ways. This is especially true for German pre-war-society where the official version of the past was closely linked to the dominating ideology. More generally such restrictions apply to any hierarchically structured local community, such as the quite conservative agrarian villages of Southern Bavaria far into the 1970s, where the teacher and the parish priest were authorities not to be challenged – at least officially.

While there is a broad scale of means of resistance against forced powers and "unquestionable" hierarchies, carnivalesque laughter is an especially subtle and widespread weapon of individuals and groups subordinate to powerful authorities (e.g. BACHTIN 1965). It is likewise a means to deconstruct and – finally – to stabilize hierarchies (TURNER 1969), in our case to question the ordered version of the past and those who construct and bring it in force, but not to revolt openly and basically accept the world as it is. In the small world of Götting it might be telling that we do not know "real" legends (except the one from barrows of Unterleiten, which is forgotten today), but stories of carnivalism against those in charge of the official version of history. They are tales of intellectual resistance and challenged authorities in times where open rejection of official narratives was beyond a villager's habitus.

One of the stories deals with the then parish priest Sebastian Knabl (1902–1926), who was one of the initiators of the Historical Society of Aibling in 1902 (STAHLER 1956: 5) and well supported in his historical research by the schoolmaster and the secretary of the post office there. He dug some of the burial mounds in the vicinity of Götting, while the farmers were draining a rivulet nearby. Suddenly they called the priest to come quickly, because they had found a well preserved iron casket. When opening the casket, Knabl found a note inside "Caesar greets his friend Reverend Knabl!" – and a barrel of beer appeared from the forest nearby to have a drink together (LOOSE 2003: 80f.). In another version of the same story Knabl was digging on the Stiegelburg and the finds he was called to dig comprised lots of old pottery – until a *Banzen* (183 litres) of stout came to light. Unfortunately the government was informed, the priest was reprimanded and had to deliver all his finds to the regional museum in Munich (LOOSE 1998: 8).[20]

In a second story the local guardian of heritage (*Heimatpfleger*) – probably Karl Braßler himself – was the aim of mockery: when building a new farmhouse in the neighbouring village of Willing a rather small skeleton of a horse was

[20] In the 1998-version Loose closes this story with the line "- there Caesar laughed!" ("- *da lachte Cäsar*!"). The line remains unexplained, but probably it alludes to the topos that history itself – here personified by Caesar – does not want to be revealed, but remains hidden to those studying her traces. If this interpretation is correct, Loose would have transferred the subversion of the villagers onto history itself.

found. The guardian thought the skeleton to be of Celtic age and published it in a lengthy article in the local newspaper. Some time later, during the carnival(!) parade in a nearby town, the skeleton of the horse was mounted on the cart of Willing taking part in the parade and the farmer reported that the horse had died twenty years ago and had been interred by himself (LOOSE 2003: 82f.).

Conditions of appropriation – mechanisms of transformation

Banal to say but not everything repeatedly taught at school or let alone published academically enters the common knowledge even of well-educated contemporaries. And even the stories of the mediators – independently of their presence in the media – are received selectively: while Karl Braßler's visions of the past are dominating the perception of history in Götting in general, a closer look, such as with the analysis of the knowledge flows around the Stiegelburg, shows that by far not everything he wrote was included in the collective memory of the villagers. Clearly not all of the (hi)stories, which remain unappropriated are rejected for hierarchical reasons or become objects of subversion. Many of them are simply neglected and/or forgotten. Obviously making truth is not enough, but the appropriation of narratives is subject to further conditions, which together shed light on the mechanisms and logics of transforming the (hi)stories in the process of appropriation.

Memory and place

My academic perspective was severely struck when I realised that the famous "Gothic" burial is almost unknown to the villagers of Götting, while they highly esteem the quite ordinary and unexplored Stiegelburg. Only later I have understood that the barrows of Unterleiten not only provide a sultan and his treasure but also a possible key to this miracle: the legend of the sultan circulated until the 1950s or a little longer, a generation after the last barrows had disappeared. But another half century later nobody knew the legend any more. Correspondingly the local disinterest in the female burial of the Migration period regards a find – which is not even very spectacular in the museum display – without place. There is no monument or site in the landscape which gives a "Here" to the burial and to which its memory could be connected – as there are no barrows left to which a sultan could be ascribed any longer. On the other hand the ramparts and ditches of the Stiegelburg are (partly) well preserved and clearly visible on the ground.

These observations neatly correspond to Katja HROBAT VIRLOGET's (in press) research on the conceptualisation of space in a Slovenian village: stereotypical folk interpretations of the past are attached to specific places in space, i.e. the landscape reflects the local past or more generally the past is an aspect of the landscape as stories of the past make places in the living world of the village. Turned into a negative perspective this might imply that there is no memory without place. If (hi)stories are the cultural form to appropriate the environment and turn it into a life world (ASSMANN 1999: 309; INGOLD 2000: 132–151) then it is rather logical that (hi)stories without places are lost, that they will disappear as there is nothing left which needs to be or can be appropriated as a life world.

On an abstract level Maurice HALBWACHS (1925: esp. chap. 4) has demonstrated that memory is localised in space and time, as such localisations are necessary conditions to order memories. Thus memory needs place – individually and collectively (ASSMANN 1997: 39, 59f.). But while such placing of memories has been widely researched recently (e.g. ASSMANN 1999: esp. chap. 2.5; cf. Pierre Nora's concept of *Les lieux de mémoire*) very little attention has been paid to the conditions, which qualify sites to become places of memory, and to the consequences for memory, if their place in the landscape disappears. The case of Götting indicates that fundamental alterations to a place like e.g. the ploughing down of the barrows of Unterleiten cuts off memories formerly connected to the site and cedes them to oblivion.

In turn this observation implies that not every site is appropriate to become a reference for memory. As neither the meadow of the now invisible barrows nor the slightly sunken site of the former gravel pit, where the "Gothic" burial had been found, acted as local centres of memories, such sites obviously need further properties which provide the potential to attach memory to them. To say the least, sites probably need perceivable physical features to potentially turn them into places.

Regarding such conditions it seems plausible that public (hi)stories – as observed at the Stiegelburg – are attached to the visible features of a monument in place instead of the finds, while academic reasoning brings the finds centre stage as witnesses of chronology, social, economic and religious interpretation – even in the case of a medieval castle where ditches and walls are preserved until today.

It is not only memories which might lose their locations but places as well can lose the memories attached to them: if I correctly derive the place name of Götting from the Goths instead of a legendary **Coto* this place name is a neat example of transforming local identity into collective knowledge and memorising it in the landscape.[21] But as identities change a place name might lose its original reference and stays in space like a functionless relict. Now it needs new (hi)stories to attach new meaning to the name and newly appropriate it (ASSMANN 1999: 308ff.). In our case the name of /Götting/ was open for new interpretations, e.g. for an academic (hi)story assuming a legendary founder in the name of the village, placing it in the centre of a para-American settler myth of expansion and colonisation. Once the new interpretation has settled in the collective memory it is hopeless (even academically) to reactivate the old meaning and nexus even if a reference – e.g. a burial of eastern Germanic tradition – comes to light.

Educational background

While matters of place may explain the need for (hi)stories, their survival or neglect, they hardly contribute to our understanding of the mechanisms whereby some narratives are highly resilient in public knowledge (Celts and Romans), some are quickly appropriated (sacrificial sites) while others are rejected ("Illyrians", Bronze Age and Urnfield culture). Strikingly the Celts and Romans are parts of the German educational background at a very broad level at least from the late 19th century onwards. With these two buzzwords Braßler's narratives met a well-established education, to which they could be connected. Quite differently the Bronze Age, the Urnfield culture and the ethnic-historical Illyrians are parts of highly specialised academic education until today. Thus a sacrificial site of those periods is a problem as its dating is meaningless due to a lack of background and thus cannot be connected to any previous public knowledge. But a sacrificial site in itself perfectly meets the established enlightenment image of prehistoric savages driven by religion. Pragmatically the sacrificial site is associated with the closest possible isle of knowledge, which are the Celts, and thus gains a position in space and time of the local history and identity. It can be appropriated to the collective knowledge as now it has meaning. In a similar way the sultan and his formidable treasure might have been buried in the barrows of Unterleiten.

The accuracy of fit to previous knowledge and established narratives not only governs which new elements are appropriated and integrated, but might also be responsible for the rejection of negative information. The Celts and the Romans

[21] Cf. ASSMANN (1999: 305) regards the place name as a sign which inscribes the place into collective memory.

are obviously fitting so well into the public (pre-)knowledge and supposedly have such an important position in the identity of the village that neither the loss of the pieces of evidence (i.e. finds) nor academic deconstructions can harm them in any way. Quite to the contrary, the Romans on the Stiegelburg seem to serve academic convictions as well as archaeologists again and again are posing the question of Roman presence at the site despite any positive indicator – and in striking contrast to the total neglect of the Celts in the academic agenda.

Taking the educational background as a main factor of selection, which elements of the past are appropriated and arranged in narratives, may point to the ordered knowledge of school as a central filter of the past. At a closer look, however, school education merely prepares a background, which provides basic ideas of "Celts" and "Romans" and a "religious past". The concretisation of these ideas, to inspire them with meaning and plausibility and to fill this background with life are functions of social groups in and far beyond school. While school certainly re-produces general features of a common habitus knitting together and structuring a society, it is the socialisation in specific groups which substantiates the individual habitus, and which provides specific orders of meaning and plausibility (BOURDIEU 1979). The (analytical construct of) "villagers of Götting" forms one such social group with specific demands on their common past.

Wrapping up the paper

By means of three examples from the village of Götting I have attempted to look more generally at academic-public-interactions. As any analysis driven by a case study the result to some degree depends on the more or less fortunate coincidences of the case under scrutiny. The case of Götting may have emphasized features which would be less important in other cases, and it certainly has its blind spots compared to the field of academic-public-interactions in general. Nevertheless it has allowed for some general conclusions.

The simple model of "academia – mediator – public", which Elisabeth Niklasson and myself had in mind, when calling for papers for our EAA-session in Helsinki (MEIER – NIKLASSON 2012: 73), surely does not work. In accordance with Michael Cremo's self-report (CREMO this volume) what we call a mediator is not just mediating between academia and the public, but frequently is doing research her-/himself with an academic background and/or in an academic format, typifying the alleged mediator among academics. And even more importantly the so called mediators do not aim at an objective transfer of knowledge *sine ira et studio* to the public, but they select and compose narratives according to their own agendas – as Diane Scherzler makes clear for the mass media (SCHERZLER this volume).

Taking my start in post-modern theory and the ethics of plurality and multiperspective constructions of group-specific (hi)stories, I have argued that narratives about the past are primarily plausible for, and bear meaning, within the social group who created the narrative. On this basis public appropriations and transformations of academic narratives are attempts to make archaeologists' narratives meaningful outside the sphere of their origin but for "the" public or – regarding the manifoldness of society – for other social groups. In this paper I have especially emphasized narratives' function in constructing group identities at the village level. Among other benefits these (hi)stories equip sites with narrative meanings, anchor them in a deep past with a flow of tradition into the present and thus turn sites into places and convert the environment into a life world (INGOLD 2000).

Considering the conditions of appropriation and especially the necessity for narratives to meet a general educational background, it is too simple a say that the public does not care about archaeologists' (hi)stories, but picks out the cherries. Quite on the contrary – also bearing in mind the great success of archaeology in exhibitions and media quotas – I would argue that there is tremendous public interest in archaeology and a constant endeavour to connect it to public knowledge – probably with the constraint that history and the ideals of education are a bourgeois project and thus bear strong social restrictions. On a local level (hi)storic appropriations are a two-fold striving to make academic narratives meaningful in the framework of local places, monuments and traditions and – as John Bintliff argues (BINTLIFF this volume: 245f.) – to connect locality to "great" and national history.

In the tradition of modernity for many meaning is still based on truth. This concept lends special weight to academic narratives, as academia for historical reasons is still regarded to have a very intimate affair with an objectified truth. On this background academics and their narrative products are attributed high social authority; they are deemed to be able to test or evaluate the truth of history. Now and then it may be public expectation that archaeologists are called to legitimize local (hi)stories, but more often it is archaeology as a set of techniques and methods, which is employed locally. It is just this idea of archaeology making truth which drove Karl Braßler and many of his fellow homeland researchers from the 19th century onwards to conduct local digs in proof of their visions of the past – which the digs mostly fulfilled as long as they were not analysed by professional archaeologists.

Such public appropriations of archaeological methods may raise severe tensions with academically trained archaeologists. Leaving aside the legal problem of private digs, such tensions usually concern interpretation. In local reception those who conducted the dig – mostly with great personal and sometimes monetary effort – usually revealed the truth about the local past. With their appropriation of archaeological, scientific methods they simultaneously have gained the authority of academia. Academic counter arguments can easily give rise to a whiff of arrogance, become objects of authoritarian critique and be rejected for anti-hierarchical, subversive reasons.

Acknowledgements

I am thankful to Nicolas Klöcker who informed me about the Stiegelburg-entry in Wikipedia and to Florian Sepp for hints on the place name of Götting. Anna Freya Schneider kindly assisted me with the search on Karl Braßler. My special thanks go to Alexander Gramsch, Elisabeth Niklasson and Karin Reichenbach for their very helpful comments on drafts of this paper and – last but not least – to John Bintliff for his careful language revision. As much as I am thankful for their help, I insist that the faults of this paper remain my own.

References

ASSMANN, ALEIDA (1999), *Erinnerungsräume. Formen und Wandlungen des kulturellen Gedächtnisses*. München.

ASSMANN, JAN (1997), *Das kulturelle Gedächtnis. Schrift, Erinnerung und politische Identität in frühen Hochkulturen*. München.

BACHTIN, MICHAIL MICHAILOWITSCH (1965*), Творчество Рабле и народная культура средневековья и Ренессанса*. Moskwa.

BÁNFFY, ESZTER (this volume), Disarmed post-socialist archaeologies? Social attitudes to interpreting the past – an interim report from Hungary. In this volume 271–293.

BEER, MATTHIAS (ed.) (2010), *Das Heimatbuch. Geschichte, Methodik, Wirkung*. Göttingen.

BIERBRAUER, VOLKER (1994), Archäologie und Geschichte der Goten vom 1.-7. Jahrhundert. Versuch einer Bilanz. *Frühmittelalterliche Studien* 28, 51–171.

BIERBRAUER, VOLKER (2004), Zur ethnischen Interpretation in der frühgeschichtlichen Archäologie. In: Walter Pohl (ed.), *Die Suche nach den Ursprüngen. Von der Bedeutung des frühen Mittelalters.* Österreichische Akademie der Wissenschaften. Philosophisch-Historische Klasse Denkschriften 322 = Forschungen zur Geschichte des Mittelalters 8. Wien. 45–84.

BINTLIFF, JOHN (this volume), Public versus professional perceptions of an invisible heritage: A Greek case study. In this volume 237–247.

BITTERAUF, THEODOR (1905), *Die Traditionen des Hochstifts Freising 1.* Quellen und Erörterungen zur Bayerischen und Deutschen Geschichte N.F. 4. München.

BÖHME, HORST W. (1988), Zur Bedeutung des spätrömischen Militärdienstes fiir die Stammesbildung der Bajuwaren. In: Hermann Dannheimer – Heinz Dopsch (eds), *Die Bajuwaren.Von Severin bis Tassilo* 488-788 = *Ausstellungskat. Rosenheim/Mattsee* 1988. s.l. 23–37.

BOURDIEU, PIERRE (1967), Postface. In: Erwin Panofsky, *Architecture gothique et pensée scolastique. Le Sens Commun.* Paris. 133–167 [I am referring to the German translation: Der Habitus als Vermittlung zwischen Struktur und Praxis. In: Pierre Bourdieu (1974), *Zur Soziologie der symbolischen Formen.* Frankfurt. 125–158].

BOURDIEU, PIERRE (1979), *La distinction. Critique sociale du jugement.* Paris.

BRASSLER, KARL (1941), Götting erzählt. Ein prähistorischer Fund in Götting. *Aiblinger Zeitung* nos. 258 and 259 of 6. and 7.11.1941.

BRASSLER, KARL (1942), Ein merowingisches Reihengräberfeld in Götting. *Aiblinger Zeitung* no. 90 of 18./19.4.1942, 3 and no. 94 of 23.4.1942, 3.

BRASSLER, KARL (1943), Auf der "Stiegelburg". *Aiblinger Zeitung* no. 66 of 19.03.1943 and no. 76 of 31.03.1943.

BRASSLER, KARL (1950), Die "Stiegelburg" – eine keltische Anlage. Neue Funde beweisen das hohe Alter dieses Burgstalls. *Aiblinger Zeitung/Wendelstein* no. 82 of 15.07.1950.

BRASSLER, KARL (1956), Die Burgställe unserer Heimat II. Die Stieglburg bei Unterstaudhausen. *Mangfall-Bote* of 17.11.1956.

BRASSLER, KARL (1957), Alt-Götting. Eine Geschichte der Gemeinde Götting an Hand der Bodenfunde und aller vorhandenen Geschichtsquellen von den

Ursiedlungen angefangen bis zum Ende des Dreißigjährigen Krieges (1648). *Der Mangfallgau* 2, 5–54.

BRASSLER, KARL (1959a), Ur-, Vor- und Frühgeschichte der Stadt und des Landkreises Bad Aibling. Zugleich ein Führer durch den Raum "Die Vorzeit des Landkreises Aibling" im Aiblinger Heimatmuseum. *Der Mangfallgau* 4, 5–88.

BRASSLER, KARL (1959b), Reihengräber im Landkreis Aibling. Sie sind Zeugen der ältesten bajuwarischen Besiedlung unserer Heimat. *Mangfallbote* no. 116 of 23.5.1959, 4.

BRATHER, SEBASTIAN (2000), Ethnische Identitäten als Konstrukte der frühgeschichtlichen Archäologie. *Germania* 78, 139–177.

CONRAD, CHRISTOPH – MARTINA KESSEL (eds) (1994), *Geschichte schreiben in der Postmoderne. Beiträge zur aktuellen Diskussion*. Stuttgart.

CREMO, MICHAEL A. (this volume), A report from a person who appropriates archaeologists' narratives for the public. In this volume 111–131.

DANNHEIMER, HERMANN – HEINZ DOPSCH (eds) (1988), *Die Bajuwaren. Von Severin bis Tassilo 488-788 = Ausstellungskat. Rosenheim/Mattsee 1988*. s.l.

FINSTERWALDER, KARL (1952), Zur Entstehung der ing-Orte um Aibling. *Das bayerische Inn-Oberland* 23, 39–44.

FOUCAULT, MICHEL (1971), *L'ordre du discours = Leçon inaugurale au Collège de France 1970*. Paris.

FOUCAULT, MICHEL (1975), *Surveiller et punir – la naissance de la prison*. Paris.

FOUCAULT, MICHEL (1977), Intervista a Michel Foucault. In: Alessandro Fontana – Pasquale Pasquino (eds), *Michel Foucault: Microfisica del potere. Interventi politici.* Nuovo Politecnico 90. Torino. 3–28.

FRIES, JANINE CLAUDIA (1995), *Vor- und frühgeschichtliche Agrartechnik auf den Britischen Inseln und dem Kontinent. Eine vergleichende Studie.* Internationale Archäologie 26. Espelkamp.

GEERTZ, CLIFFORD (1983), *Local knowledge. Further essays in interpretive anthropology*. New York.

GIESE, HELMUT (2006), *Bruckmühl und Umgebung. Ein Heimatbuch.* Bad Aibling.

GIESE, HELMUT (2008), Die Schule Götting im Wandel der Zeit. In: Klöcker – Fuchs 2008: 279–294.

GLEIRSCHER, PAUL (2002), Alpine Brandopferplätze. In: Liselotte Zemmer-Plank (ed.), *Kult der Vorzeit in den Alpen. Opfergaben – Opferplätze – Opferbrauchtum.* Schriftenreihe der Arbeitsgemeinschaft Alpenländer. Bozen. 591–634.

GOETZ, HANS-JÜRGEN (2001), *Unsichere Geschichte. Zur Theorie historischer Referentialität.* Stuttgart.

GOTTHARD, HEINRICH (1849), *Über die Ortsnamen in Oberbayern. Programm der Studienanstalt Freising vom Jahre 1849.* Freising. [I am following the reprint Freising 1884].

GRAF, MATTHIAS (1907), *Willing (Bezirksamt Bad Aibling) in Vergangenheit und Gegenwart kulturgeschichtlich dargestellt.* Bibliothek für Volks- und Heimatkunde 69. Kaufbeuren.

GRASSINGER, JOSEPH (1857), Geschichte der Pfarrei und des Marktes Aibling. *Oberbayerisches Archiv für vaterländische Geschichte* 18, 16–112, 163–298.

HALBWACHS, MAURICE (1925), *Les cadres sociaux de la mémoire. Les Travaux de l'Année Sociologique.* Paris.

HALBWACHS, MAURICE (1950), *La mémoire collective.* Paris.

HOBSBAWM, ERIC (1983), Introduction: Inventing traditions. In: Eric Hobsbawm – Terence Ranger (eds), *The invention of tradition.* Cambridge – New York. 1–14.

HODDER, IAN (1999), *The archaeological process. An introduction.* Maldon – Oxford – Victoria.

HROBAT VIRGOLET, KATJA (in press), Conceptualization of space through folklore. On the mythical and ritual significance of community limits. In: Thomas Meier – Petra Tillessen (eds), *Archaeological imaginations of religion.* Budapest.

INGOLD, TIMOTHY (2000), *The perception of the environment. Essays in livelihood, dwelling and skill.* London – New York.

KARL, RAIMUND (2010), *Macht und Ohnmacht des positivistischen Denkens. Der Positivismus in der deutschsprachigen Ur- und Frühgeschichte unter besonderer Berücksichtigung des Instituts für Ur- und Frühgeschichte der Universität Wien.* Beiträge zur Ur- und Frühgeschichte Mitteleuropas 58. Langenweisbach.

KELLER, ERWIN (1971), Ein frühvölkerwanderungszeitliches Frauengrab von Götting, Ldkr. Bad Aibling (Oberbayern). *Bayerische Vorgeschichtsblätter* 36, 168–178.

KELLER, ERWIN (ed.) (1977), *Bodendenkmäler in Bayern. Entwurf der Denkmalliste nach Art. 2 Denkmalschutzgesetz. Teil B(1): Obertägige Bodendenkmäler*. München.

KELLER, ERWIN (1987), Der nordalpine Teil der Raetia secunda im 5. Jahrhundert. *Anzeiger des Germanischen Nationalmuseums*. 77–88.

KLÖCKER, NICOLAS (2008), Die Geschichte der politischen Gemeinde Götting. In: Klöcker – Fuchs 2008: 129–189.

KLÖCKER, NICOLAS – ALOIS FUCHS (eds) (2008), *Götting. Beiträge zur Ortsgeschichte*. Götting.

KOSSACK, GEORG (1959), *Südbayern während der Hallstattzeit*. Römisch-Germanische Forschungen 24. Berlin.

LATOUR, BRUNO – STEVE WOOLGAR (1979), *Laboratory life. The construction of scientific facts*. Sage Library of Social Research 80. Beverly Hills.

LEIDORF, KLAUS – PETER ETTEL (1999), *Burgen in Bayern. 7000 Jahre Burgengeschichte im Luftbild*. Stuttgart.

LOHALM, UWE (1970), *Völkischer Radikalismus. Die Geschichte des Deutschvölkischen Schutz- und Trutz-Bundes 1919-1923*. Hamburg.

LOOSE, HELMUT (1998), *In Götting daheim. Chronik von Götting*. Bad Aibling.

LOOSE, HELMUT (2003), *Der Mond rast durch das Mangfalltal. 60 Geschichten und Anekdoten rund um Bad Aibling*. Bad Aibling.

LYOTARD, JEAN-FRANÇOIS (1979), *La condition postmoderne*. Paris.

MEIER, THOMAS (2004a), Ökosystem, Sozialstruktur und Wirtschaftsweise im mittelalterlichen Altbaiern. *Archäologisches Nachrichtenblatt* 9, 62–67.

MEIER, THOMAS (2004b), Zwischen Karpaten und Aquitanien. Das untere Mangfalltal um 400 n.Chr. In: Gabriele Graenert – Reto Marti – Andreas Motschi – Renata Windler (eds), *Hüben und Drüben – Räume und Grenzen in der Archäologie des Frühmittelalters = Festschrift Max Martin*. Archäologie und Museum 48. Liestal. 289–303.

MEIER, THOMAS (2008), Ur- und Frühgeschichte in Götting – Ein postmoderner Versuch zwischen Wissenschaft und Identität. In: Klöcker – Fuchs 2008: 15–29.

MEIER, THOMAS – ELISABETH NIKLASSON (2012), Session A16. Public appropriations of archaeologists' narratives. In: *European Association of Archaeologists 18th annual meeting – Abstracts*. Helsinki. 73–78.

MEIER, THOMAS – FELIX SCHMITT (2007), Auf der Suche nach Herren und Bauern – Parameter ältermittelalterlicher Besiedlung in Südbayern aus Sicht der Archäologie. In: John Ragnar Myking – Gertrud Thoma – Tore Iversen (eds), *Bauern zwischen Herrschaft und Genossenschaft. Peasant relations to lords and government – Scandinavia and the Alpine region, 1000–1750.* Trondheim Studies in History. Trondheim. 133–148.

MEUTER, NORBERT (2011), Geschichten erzählen, Geschichten analysieren. Das narrativistische Paradigma in den Kulturwissenschaften. In: Friedrich Jaeger – Jürgen Straub (eds), *Handbuch der Kulturwissenschaften 2. Paradigmen und Disziplinen.* Stuttgart – Weimar. 140–155.

NIKLASSON, ELISABETH (this volume), Archaeology as European added value. In this volume 49–86.

NÜNNING, ANSGAR (1999), "Verbal fictions?" Kritische Überlegungen und narratologische Alternativen zu Hayden Whites Einebnung des Gegensatzes zwischen Historiographie und Literatur. *Literaturwissenschaftliches Jahrbuch* 40, 351–380.

PLUCIENNIK, MARK (1999), Archaeological narratives and other ways of telling. *Current Anthropology* 40, 653–668.

PLUCIENNIK, MARK (2010), Is narrative necessary? *Ethnographisch-Archäologische Zeitschrift* 51, 48–63.

REINECKE, PAUL (1924), Neue Beobachtungen an süddeutschen Grabhügeln. *Germania* 8, 94.

SCHAPP, WILHELM (1953), *In Geschichten verstrickt. Zum Sein von Mensch und Ding*. Hamburg.

SCHERZLER, DIANE (this volume), "Selecting what is important for the reader": About appropriations and transformations of archaeology in the mass media. In this volume 133–143.

SCHWARZ, KLAUS (1989), *Archäologisch-topographische Studien zur Geschichte frühmittelalterlicher Fernwege und Ackerfluren im Alpenvorland zwischen Isar, Inn und Chiemsee.* Materialhefte zur Bayerischen Vorgeschichte A45. Kallmünz/Opf.

SEGER, HANS (1926), Lausitzische Kultur. In: Max Ebert (ed.), *Reallexikon der Vorgeschichte 7*. Berlin. 251–256.

SEPP, FLORIAN (2008), Die Geschichte der Pfarrei Götting von den Anfängen bis zur Gegenwart. In: Klöcker – Fuchs 2008: 191–254.

SHAPIN, STEVEN (1995), *A social history of truth. Civility and science in sventeenth-century England.* Chicago – London.

STAHLER, FRIEDRICH WILHELM (1956), Abriß der Geschichte des Historischen Vereins für Bad Aibling und Umgebung und seines Museum. *Der Mangfallgau* 1, 5–12.

STEFFAN, FERDINAND –HANS PETER UENZE (2003), *Vor- und Frühgeschichte in Stadt und Landkreis Rosenheim II.* Kataloge der Archäologischen Staatssammlung München 29. München.

TUAN, YI-FU (1977), *Space and place. The perspective of experience.* Minneapolis.

TURNER, VICTOR (1969), *The ritual process. Structure and anti-structure.* New York.

VEIT, ULRICH (2010), Zur Geschichte und Theorie des Erzählens in der Archäologie: Eine Problemskizze. *Ethnographisch-Archäologische Zeitschrift* 51.1/2, 10–29.

WALLNER, EDUARD (1950), Beiträge zum Namensregister der Traditionen des Hochstifts Freising, herausgegeben von Theodor Bitterauf. *Oberbayerisches Archiv für vaterländische Geschichte* 76, 43–79.

WEBER, FRANZ (1886), Ausgrabungen vorgeschichtlicher Alterthümer in Baiern im II. Halbjahr 1886. Hügelgräber. *Mittheilungen des Museums-Vereins für vorgeschichtliche Alterthümer Baierns* 9, 6–8.

WEBER, FRANZ (1909), *Die vorgeschichtlichen Denkmale des Königreiches Bayern 1. Oberbayern.* München.

WEISS, RAINER MARIA (1997), *Prähistorische Brandopferplätze in Bayern.* Internationale Archäologie 35. Espelkamp.

WELSCH, WOLFGANG (1991), Postmoderne – Pluralität als ethischer und politischer Wert. In: Jörg Albertz (ed.), *Aufklärung und Postmoderne – 200 Jahre nach der französischen Revolution das Ende der Aufklärung?* Schriftenreihe Freie Akademie 11. Berlin. 9–44.

WHITE, HAYDEN (1973), *Metahistory. The historical imagination in nineteenth-century Europe.* Baltimore.

WHITE, HAYDEN (1987), *The content of the form. Narrative discourse and historical representation.* Baltimore.

Neopaganism, archaeological content and the belief in "Celts"

JUTTA LESKOVAR

Abstract:

The question to be pursued in this paper is how and to what extent archaeological terms, objects, places, etc. are used by some authors of Neopagan books (Neodruidism and Wicca). Those examined regularly cite archaeological publications, which make it possible to track the way archaeological content, in this case mainly about 'the Celts', took from a scholarly context into the Neopagan world. It also allows us to examine the scholarly sources for Neopagan interpretations. And it is these sources that have proven to be extremely useful for Neopagan purposes: in them, the term 'Celts' is used for places and topics that should not be connected with it; as a generic term to describe Iron Age Europe. Even more importantly, the 'source archaeologists' tend to describe prehistoric religion in ways that are reminiscent of Neopagan texts, making them even more useful for authors of Neopagan literature.

Neopaganism and its use of archaeological arguments for constituting and describing its worldview and religion has as of yet been widely regarded within archaeology as an inappropriate misuse and falsification of archaeological knowledge. Many strands of Neopaganism rely heavily on information about the past, the past being their main source for (re-) creating a religion. Using archaeological publications does not necessarily mean to adopt wholesale their scholarly results and conclusions, like chronological systems or the dating of any given site. Nor are Neopagan writers hesitant to create whole religious systems for which hardly any (archaeological or other) evidence exists.

This paper aims at giving an introduction about Neopaganism as a modern phenomenon (for general information see [Davy 2007]), to be followed by a discussion about how Celtic Neopaganism uses archaeological facts as arguments for its beliefs, and how the way archaeologists use the word "Celts" influences what non-scientists write about the past. It will conclude with a few general remarks about the relationship between Neopaganism and Archaeology.[1]

[1] The results of this paper are based on my PhD thesis "Kämpfen um die Kelten" (Leskovar 2012c). Several German articles have already been published (Leskovar 2009; 2010; 2012a; 2012b; in press).

Neopaganism

All parts of the Neopagan world draw heavily on the past or on certain ways to believe in the past (LESKOVAR 2012c: 42f.). Neopaganism is not only a scientific term to describe a modern religious movement, but also a self-designation of its members. To be a Pagan means, to put it rather bluntly, to be opposed to Christianity or any other monotheistic religion (LESKOVAR 2012c: 139–141), to look back frequently to the past, mainly to certain periods before Christianity. Neopaganism, as expressed by the word itself, aims at creating something new, while staying in a pagan, pre-Christian tradition. The feeling of living and practicing a genuinely old, even prehistoric religion, seems to be especially strong (TANITA-PAN 2002: 16–30; CARR-GOMM 2004: 19–36). "Recreation" is, in this context, a typical word, used by Pagans themselves (STARHAWK 1988: 21). It seems to be an expression of a certain awareness that much is lost, which causes the need to re-create, to fill the huge gaps between the pieces that have survived. Obviously the question about what has survived is one of the crucial points in the discussion between Neopagans and Archaeologists about the prehistoric past.

If queried, Archaeologists would normally express the opinion that what actually *has* survived is so little that no one should start to use it as a base for (re-)creating a religion. But, since they think it is possible, this is exactly what Neopagans do.

It is difficult to describe the Neopagan scene in short words, as it is, in many aspects, extremely heterogeneous. No church exists, no clergy that is accepted by all Neopagans, no holy book, etc. Neopagans are people who practice a religious tradition that is about 60-100 years old,[2] at least from a historian´s point of view (HUTTON 2001; 2007; 2009), while thousands of years old in the mind of Neopagans themselves. Many Neopagan strands call themselves after an ethnic group: Egyptian Neopagans, Celtic, Greek, Roman, German, Slavic, Baltic, Finnish Neopagans, and many more of this kind exist. Another, rather big group are the so called Witches, many of which can be connected with the term "Wicca". To understand Neopagan beliefs it seems to be helpful to state a few details about what they actually do.

---

[2] The first druidic orders go back to the 18th century. Their main focus was, at least at the beginning, not practising esoteric or neopagan traditions, but rather charitable work (HUTTON 2007: 137–155).

The importance of a connection between humans and the natural world is central to Neopagan thinking (LESKOVAR 2012c: 42; CARR-GOMM 2004: 19; CUNNINGHAM 2001: 24). Our modern technological world is described as having cut the original link between humans and nature. Neopagan life should be able to establish this connection again, give us back roots and give nature back its value. Therefore, outdoor celebrations, sometimes even at archaeological sites – Stonehenge being the most prominent example – are a typical, although no strictly necessary feature of the Neopagan world. Celebrations, in groups of different sizes or alone, are a central point in Neopagan thinking and practising (LESKOVAR 2012c: 43). Neopagans, on the whole, accept and practice a festival circle around the year, even if names, content and rituals differ (CARR-GOMM 2004: 111–124; CONWAY 2001: 273–286; DAVY 2007: 56; PAXSON 2006: 109–113).

One of the "classical" Neopagan calendars includes eight seasonal festivals: the so called "Celtic" festivals – best known examples are Beltane at the beginning of May and Samhain at the beginning of November, which takes place at the same date as Halloween or All Saints Day (the others are Imbolc and Lughnasad) – in combination with four astronomically marked festivals, which are celebrated at summer/winter solstice and the equinoxes. Festivals or celebrations always include rituals which demand the usage of certain objects. The variety is huge, but it may include knifes, daggers, cauldrons etc. (TANITA-PAN 2002: 84, 88; STARHAWK 1988: 89, 108, 111) – objects that are also known from archaeological contexts.[3]

Believing in the great, even prehistoric age of seasonal festivals is one of the pillars of Neopagan creed. This conclusion is drawn mainly from medieval Irish sources, which name and describe, although in just a few words, certain festivals (KELLY 2000: 460–462). Obviously, those sources cannot prove a prehistoric tradition of the festivals, and certainly not for the entirety of Europe.

Neopagans do not, among themselves, have the same view of the past, or name the same periods as important for the development of their certain traditions. The pattern how archaeology is used is fairly similar within the different branches. From a scientific viewpoint we are, in Europe, currently best informed about Celtic and German Neopaganism and Wicca.

---

3 There has so far no research been carried out concerning the comparison of archaeological types and objects used in Neopagan contexts.

Wicca, German and Celtic neopaganism

Wicca is best described as the religion of modern witches. It was established by Gerald Gardner in the 1940s in Southern Britain (see HUTTON 2001). Gardner, whose life and work is well recorded, mixed a lot of mythological and occult strands of his time, to create Wicca as a modern Neopagan religion. The aspect of recreating something old, which has survived below the surface during centuries of Christianity, which could be pulled to light again, was very strong from the very beginning of this creative act. Gardner's early co-founders emphasised this belief. There even have been scholars, working during his lifetime, who supported those ideas. Margaret Murray, an acclaimed Egyptologist, published her theory about the old religion of the witches, who survived until the so called "Burning Times" of the early modern period, in 1931 (MURRAY 1931). Wiccans frequently refer to those "Burning Times", believing to be heirs to the victims of the witch trials, who, in their view, have been protagonists of a true Pagan religion. This seems to be a dangerous view, as it proclaims the victims to be guilty of the charge retrospectively, even centuries later.

Wiccans have adopted shamanic ideas and practices, which gives them the chance to take every clue of prehistoric shamanism as proof of the old age of their own religion: as Wicca is shamanic, every old religion which is interpreted as (in parts) shamanic can be regarded as a forebear of Wicca. Palaeolithic drawings in caves are interpreted as signs of early shamanism – by archaeologists, too (HAHN 1994: 114; NEUGEBAUER-MARESCH 1993: 18; STREET 1989). Therefore, archaeological publications can be used as valuable source by Wiccans who write the first chapter of a book about the Craft, which nearly always includes an alleged line of tradition from the Stone Age to the witches of the witch trials and up to the revival, led by Gerald Gardner.

German Neopaganism has been subject to comparatively many studies by German speaking scholars of different professions (DUSSE 2006; MÖLDERS – HOPPADIETZ 2007; PÖHLMANN 2006; VON SCHNURBEIN 1992a; 1992b; 1996a; 1996b; 1996c; 2001; 2004). It is also known by the term "Asatru", which means "True to the Ases", the high gods of Norse mythology. German Neopagans use, beside other sources, the Edda for recreating their beliefs and rituals. The scene has often been accused of propagating nationalistic and racist views in the tradition of the Nazis. Many groups of the German Neopagan scene can rightly be accused of this, others not. It is necessary to look very closely at single groups and even individual members to draw conclusions of this kind.

Celtic Neopagans have others in mind when trying to establish their line of tradition – "Celts" and "Druids" are their chosen ancestors. Many Celtic Neopagans are organised in Druidic Orders, some of which already exist for decades, mainly in the UK. The greatest current Druidic Neopagan order is the Order of Bards Ovates and Druids, which acts worldwide. The publications of its leader, the so called Chosen Chief, Philip Carr-Gomm, are fine examples of how archaeology is used in Celtic Neopaganism (CARR-GOMM 2004).

Celtic neopaganism and archaeology

In order to better understand this process of reinvention (of prehistoric religions), I analysed 10 publications which describe Celtic Neopaganism, or Druidry, in a general way (LESKOVAR 2012c). I aimed at filtering to what extent archaeological results or topics are used as arguments for the great age of Neopagan beliefs. Summing up the results, archaeology is used constantly but not in large amounts. Archaeological results may be cited just five times in a book of three hundred pages, but they will appear in certain circumstances: whenever the author needs some kind of connection to the past. Archaeological facts act as evidence that what the author says is true, because it is old.

Neopagans tend to use linguistic results, or ancient authors, or Irish or Welsh literature more often than archaeological results. Written sources seem to guarantee more liveliness, give the story about an ancient culture and its religion more feeling than mere objects and sites, analysed and described by archaeology. Nevertheless, archaeology has its value too, in providing Neopagans with hard facts, with seemingly true scientific information, from their point of view. Of course, the problem with archaeology, and science on the whole, is that it seldom delivers any "true" scientific information. There are sites and objects, and more often than not less than perfect interpretations. This characteristic of archaeological evidence makes it even more useful to create an interpretation of one´s own.

What is important about the prehistoric past in a Neopagan author´s view?

As already stated above, Neopagans reach as far back as the Palaeolithic, always in connection with shamanism. Next in chronological line are the Megaliths. Some authors point out that Stonehenge was not erected by the Druids, while at the same time trying to connect the wisdom of building a Megalithic site with the (Iron Age) Druids. Carr-Gomm, for example, solves this problem by speaking about "proto-Druids" as creators of Stonehenge (CARR-GOMM 2004: 29–31).

The descriptions and interpretations of Stonehenge and other Megaliths are fine examples of the typical ways archaeological facts are used by Neopagans. Neopagan authors are perfectly able to give a precise description of any given archaeological site, combined with a correct dating, but they still create their own interpretations around it.

Books about Druidry nearly always contain an introduction about the origin of "Celtic culture" (BENZ – SCHREGER 2002: 26–29; BÖCKL 2004: 16–25; CARR-GOMM 2004: 24–34; COWAN 1998: 13f.; EDEL – WALLRATH 2000: 8f.; KALWEIT 2002: 36–40; RABEY 2002: 12). While trying to describe a consistent chronological line of development, Neopagans often use terms like Bell Beaker culture, Urnfield culture, and, in this case, of course Hallstatt culture, Latène culture, etc. Neopagans very often describe certain sites, mainly from the British Isles or from Germany, depending on the origin of the author (LESKOVAR 2012c: 129–134). British hillforts, or sites like Hochdorf, Mont Lassois or the Heuneburg are used to paint a certain picture of Iron Age culture, mainly of the society, which is often described as tribal and "natural". In some cases sites are also used to indicate the high social status of Celtic women (EDEL – WALLRATH 2000: 112f., 126; O´DONOHUE 1999: 215f.). Late Iron Age "*Viereckschanzen*" are obviously interpreted as cult sites, a fact that does not surprise, as the discussion about their purpose is not finished yet and some archaeologists tend towards this interpretation too (LESKOVAR 2012b: 383). Remarks about sites like Roquepertuse are made frequently, combined with an alleged importance of heads in the Celtic world (LESKOVAR 2012b: 383). Certain objects are named, not very often and mainly in connection with the typical Latène style, which is interpreted in an alleged mythological way of Celtic thinking. The Gundestrup cauldron is extremely well known and described very often (LESKOVAR 2012b: 384; BENZ – SCHREGER 2002: 125; BÖCKL 2004: 75; COWAN 1998: 267; KALWEIT 2002: 349; RABEY 2002: 267). It carries a lot of scenes and figures that are open for interpretations of any kind, and by anybody, be it Neopagans or Archaeologists. This Danish cauldron, and some others, for example the one of Hochdorf, bear witness, at least in Neopagan´s opinions, that cauldrons were of big ritual importance in the Iron Age.

Still, something is missing. Interestingly, burials are never mentioned by Neopagan authors, or only in context with certain objects (LESKOVAR 2012b: 384–386). This seems strange, as especially archaeologists look closely at burials, for lack of other sources (except hoards), for answering questions about mythological or religious views in prehistory. It seems that the sphere of life is much more interesting for Neopagans than the sphere of death, mythological or not.

It has been shown that Neopagans use a lot of different archaeological themes (LESKOVAR 2012c). Their knowledge is obviously gained by reading archaeological publications. Most of the Neopagan books which have been analysed included a bibliography, the bigger part of which consisted of books from the period after 1990. General descriptions of Celtic culture by archaeologists or by scholars of religious studies or linguistics were included, if to a lesser extent than books about Neopagan or other esoteric/new age topics. None of the Neopagan authors used any kind of citation in the text itself, so it is impossible to trace exactly where which piece of information was taken from. However, I proceed from assuming that the archaeological publications which are part of the bibliographies were indeed read, or at least looked at. This led to another analysis of a part of the scholarly books which are part of the Neopagan bibliographies (BIRKHAN 1997; BITTEL ET AL. 1990; CUNLIFFE 1980; DANNHEIMER – GEBHARD 1993; DEMANDT 1998; KRUTA – SZABÓ 1979; MAIER 2001; PAULI 1975; PIGGOTT 1993; RAFTERY 1994), which aimed at answering the following question: Which definition of "Celts" and "Celtic culture" is used by the authors of the scholarly publications (LESKOVAR 2012c)? The way archaeologists or other scholars use the term "Celt" for a lot of different things could be part of the reason why there are so many "strange" Neopagan opinions about the past – at least this was part of the assumption before starting the analysis. Part of this assumption could be proven during the analysis (LESKOVAR 2012c). Definitions of "Celts" are only rarely or never found in those scholarly publications, which have been used by the authors of those Neopagan books that had been selected for the first analysis. And even if a definition is given, it is, more often than not, not used consistently throughout the book. For example, if "Celts" have been defined as "people who spoke a Celtic language", archaeological finds from Ireland and southern Germany are used to illustrate Celtic culture as well (BIRKHAN 1997: 45; MAIER 2001: 11ff.). No consistent chronological pattern exists. "Celtic" is used as a synonym for "La Tène period" on a regular basis (BIRKHAN 1997: 26f., 32, 46; BITTEL ET AL.

1990; DEMANDT 1998: 14; DANNHEIMER – GEBHARD 1993: 2; PAULI 1975: 191, 205; PIGGOTT 1993: 26). It is also often used synonymously for "Hallstatt period" (BIRKHAN 1997: 45; DEMANDT 1998: 15f.; KRUTA – SZABÓ 1979: 26ff.; MAIER 2001: 11ff.; RAFTERY 1994: 10; PAULI 1975: 14) and sometimes even "Urnfield period" (CUNLIFFE 1980: 7; PIGGOTT 1993: 25ff.). This broad view reappears in a geographical sense: Ireland as the western and Anatolia as the eastern boundary create a big Celtic realm which covers the better part of Europe.

As Neopagans are bound to be interested in what scholars write about "Celtic religion", it seemed useful to make another analysis about the way scholars describe "Celtic religion" in their books (LESKOVAR 2012a). And it is, with a few exceptions, almost always described in a way that reminds the reader more of the language of Neopagans than scholarly writing. Wolfgang Kimmig and Barry Cunliffe give two good examples:

> "The Celt was searching for and found the workings of the divine in springs and sacred groves, in the whispering of trees in the wind, in oracles and mysterious cults, which probably were celebrated in these nemeta. Especially tree cults must have played a special role in these" (KIMMIG 1993: 172).

> "The Celts were a superstitious people. The supernatural pervaded all aspects of their lives – the spirits were everywhere: in old trees, mysterious rocks, in rivers and bogs. Even the most ordinary tasks were subject to the influence of the gods. They ruled the natural world, of which man was part: that is why they had to be placated by arbiters – the druids. As those preserving old wisdoms, they had to ensure that the right procedures were always followed when interacting with the gods" (CUNLIFFE 1980: 69).[4]

In recreating a picture of a mythological and long gone past, full of cult and religious feeling, Neopagans and archaeologist seem to have something in common at last. What is unclear, is whether one received this kind of language from the other in the first place, and which way a possible transfer took.

---

[4] I would like to thank Raimund Karl for translating the German passages.

Conclusions

On the whole, authors of Neopagan books in search of information about "the Celts" will find a lot of different and very useful pieces of information in a lot of books written by scholars. The use of the term "Celt" for very different topics from different times and regions, often centuries and thousands of miles apart, must give non-specialists the feeling that their view of a certain people, occupying a certain space in history, is correct. In carrying this thought further, this gives them the assurance that it makes sense to use whatever they find in other books for their purpose, as long as the word "Celtic" sticks to it. This process will, of course, produce results that have not been approved of by the original writers of the scientific publications, but they have, involuntarily, helped to create those views.

What can be done about this? I do not believe that a change in using the word "Celt" in academia will change anything of what is written about the past in Neopagan books. I just believe it is important, as an archaeologist, to know about those connections. Nonetheless, it seems necessary to carry on with discussions about the usefulness of ethnic terms, be it "Celts" or "Germans", in archaeological contexts and publications (COLLIS 2003; KARL 2004; RIECKHOFF 2007; JARNUT 2004). Definitions have to be followed, at least by those who formulate and use them in their own books. The problems with ethnic terms must be repeated as often as they are used – not to stop the public using the terms in ways archaeologists may not like, but to inform the interested part of the public about current scholarly debates. To simplify the past is a privilege of the layman – for example Neopagans writing about their view of the "Celtic past". Archaeologists are bound to explain the difficulties and variety of interpretations about the past.

Neopagans are often accused not to write in a scholarly way, not to use correct citations, or just to use scholarly results of others in the first place. On the whole, they are not to blame, as long as they do not pretend to *be* scholars, and they are not forced to act and write in the way we, as part of a scholarly community, feel obliged to do. Neopagans are certainly allowed to use published material as any other person, scholar or not. However, it still may be a nuisance to watch Neopagan authors use scholarly results in a non-scholarly way, as arguments for their, from an archaeological point of view often strange, beliefs. Neopagans use the commonly accepted value of scholarly knowledge for their own purpose – if you claim cauldrons to be of great importance for prehistoric religions, it makes sense to name prehistoric burials with cauldrons in it, if you want to convince your reader. In this way, Neopaganism relies heavily on scholarship, on archaeology, but it neither uses our ways, nor does it always appreciate the importance of scholarly work. On the contrary, scholarship and scholars are often blamed for holding back information that could support Neopagan ideas about continuity of religious practices and beliefs. Scholars are blamed for their results – one striking example is the discussion after Ronald Hutton published his study about the very recent (first part of the 20th century) development of Wiccan religion (HUTTON 2001). In parts, there was uproar in the Wiccan world, and Hutton was blamed for trying to destroy their beliefs. I have even come across the Neopagan opinion that scholars should not be allowed to analyse what is a main topic for Neopagans – namely prehistoric religions (see TULLY 2011: 99f.). It may be regarded as a strange view, but it is nothing but the correlate of some scholars' opinion – that non-scholars should not be allowed to write about scholarly results about the past.

The relationship between Neopaganism and the world of scholarship is still complicated, sometimes messed up. Nothing but knowledge about each other and dialogue can get us anywhere. One of the most important reasons why archaeologists should be acquainted with the Neopagan scene is the non-Neopagan public. Many people just want to know something about the past, without sharing Neopagan interests. Those people have a much greater chance of coming across a Neopagan book about Celtic religion in their bookstore on the internet, than a scholarly study of the same topic. Therefore it is our task to keep the public informed about the Neopagan phenomenon from an archaeologist´s viewpoint. We need to tell the public, not that Neopaganism is bad, but that its claims for practicing an old tradition are false. We should also not forget that a lot of Neopagans are interested in what is and what is not correct about the beliefs about the past of those who founded their religion.

References

BENZ, DORIS – BEN SCHREGER (2002), *Kelten, Kulte, Anderswelten. Auf Spurensuche. Schwarzwald – Elsass – Schwäbische Alb – Oberschwaben.* Unterweitersdorf.

BIRKHAN, HELMUT (1997), *Kelten. Versuch einer Gesamtdarstellung ihrer Kultur*. Wien.

BITTEL, KURT – SIEGWALT SCHIEK – DIETER MÜLLER (1990), *Die keltischen Viereckschanzen.* Atlas archäologischer Geländedenkmäler in Baden-Württemberg 1. Stuttgart.

BÖCKL, MANFRED (2004), *Die Botschaft der Druiden. Weisheit aus der Anderswelt*. Saarbrücken.

CARR-GOMM, PHILIPP (2004), *Die Weisheit der Druiden. Eine Einführung in die keltische Spiritualität*. Stuttgart.

COLLIS, JOHN (2003), *The Celts. Origins, myths and inventions*. Stroud.

CONWAY, DEANNA J. (2001), *Wicca. The complete craft*. New York.

COWAN, TOM (1998), *Die Schamanen von Avalon. Reisen in die Anderswelt der Kelten*. Kreuzlingen – München.

CUNLIFFE, BARRY (1980), *Die Kelten und ihre Geschichte*. Bergisch Gladbach.

CUNNINGHAM, SCOTT (2001), *Wicca. Eine Einführung in weiße Magie*. München.

DANNHEIMER, HERMANN – RUPERT GEBHARD (eds) (1993), *Das keltische Jahrtausend*. Mainz.

DAVY, BARBARA JANE (2007), *Introduction to pagan studies*. Lanham – Plymouth.

DEMANDT, ALEXANDER (1998), *Die Kelten*. München.

DUSSE, DEBORA (2006), Eine moderne Religion in alten Zeiten. Germanische Religionsgeschichte und Neuheidentum. In: Matthias Pöhlmann (ed.), *Odins Erben. Neugermanisches Heidentum. Analyse und Kritik.* Evangelische Zentralstelle für Weltanschauungsfragen - Texte 184. Berlin. 37–50.

EDEL, MOMO – BERTRAM WALLRATH (2000), *Götter, Barden & Druiden. Die Kelten. Europas spirituelle Kindheit*. München.

HAHN, JOACHIM (1994), Menschtier- und Phantasiewesen. In: Gerhard Bosinski – Brigitte Reinhardt (eds), *Der Löwenmensch. Tier und Mensch in der Kunst der Eiszeit*. Begleitpublikation zur Ausstellung Ulm. Ulm. 101–115.

HUTTON, RONALD (2001), *The triumph of the moon. A history of modern pagan witchcraft*. Oxford.

HUTTON, RONALD (2007), *The Druids*. London.

HUTTON, RONALD (2009), *Blood and mistletoe. The history of the druids in Britain*. New Haven – London.

JARNUT, JÖRG (2004), Germanisch. Plädoyer für die Abschaffung eines obsoleten Zentralbegriffes der Frühmittelalterforschung. In: Walter Pohl (ed.), *Die*

Suche nach den Ursprüngen. Von der Bedeutung des frühen Mittelalters. Forschungen zur Geschichte des Mittelalters 8. Wien. 108–113.

KALWEIT, HOLGER (2002), *Das Totenbuch der Kelten. Das Bündnis zwischen Anderswelt und Erde*. Aarau.

KARL, RAIMUND (2004), Die Kelten gab es nie. Sinn und Unsinn des Kulturbegriffs in Archäologie und Keltologie. In: Raimund Karl (ed.), *Archäologische Theorie in Österreich. Eine Standortbestimmung*. Wien. 7–35.

KELLY, FERGUS (2000), *Early Irish farming*. Dublin.

KIMMIG, WOLFGANG (1993), Menschen, Götter und Dämonen – Zeugnisse keltischer Religionsausübung. In: Dannheimer – Gebhard (1993), 170–176.

KRUTA, VENCESLAS – MIKLÓS SZABÓ (1979), *Die Kelten. Entwicklung und Geschichte einer europäischen Kultur in Bildern von Erich Lessing*. Freiburg im Breisgau.

LESKOVAR, JUTTA (2009), Uraltes Neuheidentum? Archäologische Argumente in der "esoterischen" Literatur und ihre Herkunft. In: Christiana Eggl – Peter Trebsche – Ines Balzer – Janine Fries-Knoblach – Julia K. Koch – Hans Nortmann – Julian Wiethold (eds), *Ritus und Religion der Eisenzeit. Beiträge zur Sitzung der AG Eisenzeit während der Jahrestagung des Mittel- und Ostdeutschen Verbandes für Altertumsforschung e.V. in Halle an der Saale 2007.* Beiträge zur Ur- und Frühgeschichte Mitteleuropas 49. Weissbach. 19–27.

LESKOVAR, JUTTA (2010), Alte Kelten – Neue Druiden. Archäologie, Neuheidentum und der Keltenbegriff. *Oberösterreichische Heimatblätter* 64.3/4, 119–131.

LESKOVAR, JUTTA (2012a), "Keltische Religion". Neuheidnische und wissenschaftliche Sichtweisen. In: Raimund Karl – Jutta Leskovar – Stefan Moser (eds), *Die erfundenen Kelten – Mythologie eines Begriffes und seine Verwendung in Archäologie, Tourismus und Esoterik. Interpretierte Eisenzeiten. Fallstudien, Methoden, Theorie. Tagungsbeiträge der 4. Linzer Gespräche zur interpretativen Eisenzeitarchäologie – Tagung im Keltenmuseum Hallein Nov. 2010.* Studien zur Kulturgeschichte von Oberösterreich 31. Linz. 239–255.

LESKOVAR, JUTTA (2012b), Eisenzeitliche Argumente im Neodruidentum. In: Peter Anreiter – Eszter Bánffy – László Bartosiewicz – Wolfgang Meid – Carola Metzner-Nebelsick (eds), *Archaeological, cultural and linguistic heritage. Festschrift for Erzsébet Jerem in honour of her 70th birthday.* Budapest. 379–391.

LESKOVAR, JUTTA (2012c), *Kämpfen um die Kelten. Archäologische Argumente in der neuheidnischen Literatur und der Keltenbegriff in der Fachliteratur.* Keltische Forschungen, Allgemeine Buchreihe 2. Wien.

LESKOVAR, JUTTA (in press), "Brigid, I call thy name" – Keltische Göttinnen im Neuheidentum. In: Julia K. Koch – Christina Jacob (eds), *Prähistorische und Antike Göttinnen. Befunde – Interpretation – Rezeption. Jubiläumstagung des FemArc-Netzwerk archäologisch arbeitender Frauen e.V., März 2011.* Heilbronn.

MAIER, BERNHARD (2001), *Die Religion der Kelten. Götter, Mythen, Weltbild.* München.

MÖLDERS, DOREEN – RALF HOPPADIETZ (2007), "Odin statt Jesus!" Europäische Ur- und Frühgeschichte als Fundgrube für religiöse Mythen neugermanischen Heidentums? *Rundbrief Theorie AG* 6/1, 32–48.

MURRAY, MARGARET (1931), *The god of the witches.* London.

NEUGEBAUER-MARESCH, CHRISTINE (1993), Zur altsteinzeitlichen Besiedlungsgeschichte des Galgenberges von Stratzing/Krems-Rehberg. *Archäologie Österreichs* 4/1, 10–19.

O´DONOHUE, JOHN(1999), *Anam Ċara. Das Buch der keltischen Weisheit*[6]. München.

PAULI, LUDWIG (1975), *Keltischer Volksglaube. Amulette und Sonderbestattungen am Dürrnberg bei Hallein und im eisenzeitlichen Mitteleuropa.* Münchner Beiträge zur Vor- und Frühgeschichte 28. München.

PAXSON, DIANA L. (2006), *Essential Ásatrú. Walking the path of Norse paganism.* New York.

PIGGOTT, STUART (1993), *The druids.* London [first published 1968].

PÖHLMANN, MATTHIAS (2006), Rückkehr zu Odin und Freyja? Neugermanisches Heidentum – eine evangelische Kritik. In: Matthias Pöhlmann (ed.), *Odins Erben. Neugermanisches Heidentum. Analyse und Kritik.* Evangelische Zentralstelle für Weltanschauungsfragen – Texte 184. Berlin. 68–96.

RABEY, STEVE (2002), *Das Wissen der Kelten.* Düsseldorf.

RAFTERY, BARRY (1994), *Pagan Celtic Ireland.* London.

RIECKHOFF, SABINE (2007), Die Erfindung der Kelten. In: Raimund Karl – Jutta Leskovar (eds), *Interpretierte Eisenzeiten. Fallstudien, Methoden, Theorie. Tagungsbeiträge der 2. Linzer Gespräche zur interpretativen Eisenzeitarchäologie.* Studien zur Kulturgeschichte von Oberösterreich 19. Linz. 23–37.

STARHAWK(1988), *Der Hexenkult als Ur-Religion der Großen Göttin. Magische Übungen, Rituale und Anrufungen*[4]. Freiburg im Breisgau.

STREET, MARTIN (1989), *Jäger und Schamanen. Bedburg-Königshoven. Ein Wohnplatz am Niederrhein vor 10.000 Jahren*. Mainz.

TANITA-PAN, COVEN (2002), *Hexen des Alten Weges. Praktische Magie und die Mysterien von Transzendenz und Macht*. München.

TULLY, CAROLINE JANE (2011), Researching the past is a foreign country: Cognitive dissonance as a response by practitioner pagans to academic research on the history of pagan religions. *The Pomegranate* 13/1, 98–105.

VON SCHNURBEIN, STEFANIE (1992a), *Göttertrost in Wendezeiten. Neugermanisches Heidentum zwischen New Age und Rechtsradikalismus*. München.

VON SCHNURBEIN, STEFANIE (1992b), *Religion als Kulturkritik. Neugermanisches Heidentum im 20. Jahrhundert*. Heidelberg.

VON SCHNURBEIN, STEFANIE (1996a), Die Suche nach einer 'arteigenen' Religion in "germanisch"- und "deutschgläubigen" Gruppen. In: Uwe Puschner – Walter Schmitz – Justus H. Ulbricht (eds), *Handbuch zur "Völkischen Bewegung" 1871–1918*. München. 172–185.

VON SCHNURBEIN, STEFANIE (1996b), Weiblichkeitskonzeptionen im neugermanischen Heidentum und in der feministischen Spiritualität. In: Otto Bischofberger – Peter Hölzle – Stefanie von Schnurbein (eds), *Das neue Heidentum. Rückkehr zu den alten Göttern oder neue Heilsbotschaft?* Freiburg (Schweiz). 42–71.

VON SCHNURBEIN, STEFANIE (1996c), Neuheidnische Religionsentwürfe von Frauen. In: Otto Bischofberger – Peter Hölzle – Stefanie von Schnurbein (eds), *Das neue Heidentum. Rückkehr zu den alten Göttern oder neue Heilsbotschaft?* Freiburg (Schweiz). 72–103.

VON SCHNURBEIN, STEFANIE (2001), Transformationen völkischer Religion seit 1945. In: Stefanie von Schnurbein – Justus H. Ulbricht (eds), *Völkische Religion und Krisen der Moderne. Entwürfe "arteigener" Glaubenssysteme seit der Jahrhundertwende*. Würzburg. 409–429.

VON SCHNURBEIN, STEFANIE (2004), Religion of nature or racist cult? Contemporary Neogermanic pagan movements in Germany. In: Hubert Cancik – Uwe Puschner (eds), *Antisemitismus, Paganismus, Völkische Religion.* München. 135–149.

The Acropolis of Athens as imaginary neighbour in the local 'homeland'

STELLA KATSAROU-TZEVELEKI

Abstract:
The Acropolis, a national and global symbol constituted through mutual colonial engagements with western modernity, in addition to its non-reflexive image of immortality, also bears hidden faces behind its solidified austerity. Faces like those generated within the range of contingent individual discourses occurring between the monument and the society of permanent residents in the area around the Sacred Rock. Amongst these residents, the pupils of the very local Primary School have subscribed their personal appropriations of the monument to the official claim of the Greek state for the return of the Parthenon marbles to Athens. They have done so by particularising their argument based on their personal trauma, suffered from the exile of the violently dismembered ancient properties out of their native land. In the course of a locally initiated project, run outside of the state's official campaign, their claim has been propagated by an anthology of artworks, citing imaginary versions of the schoolchildren's memories. The images show the presences and absences of the ancient sculptures from their neighbourhood, advocated as the very powerful homeland. The present paper discusses the project by stressing its indigenous voice upon the long lived dream of antiquity residing at the core of Hellenic nationhood, and the response of these accounts to post-modern multiple contexts.

The 'young neighbours of the Acropolis' project

Konstantinos on top of a marble horse from the Parthenon frieze is riding over the hills in front of the Acropolis. At the playground the 'mourning goddess Athena' is looking at Nefeli who's cuddling a Caryatid, and further behind her another Caryatid is sitting on a swing.

These are two out of the 82 images created by the Acropolis' schoolchildren, stating their imaginary and metaphysic biographies with the monument's sculptures, narrated to respond to a locally managed project. The project aims to promote the bring-them-back appeal for the Parthenon marbles as essential to the children's constant embodied appropriation of the monumental artefacts, as well as their emotional bondedness with the sculpted images. As pupils of the 70th Primary School of Athens, built at the foot of the Acropolis hill to its south and across the street on the west side of the new Acropolis Museum, these children have indeed been spending their schooldays in a constant visual dialogue with the pinnacle of Greece's emblematic past. The appearance of the Acropolis rock, and in particular of the Parthenon, dominates the view from the school courtyard (*Fig. 1*). The penetrating view that one can have from the year 5 and 6 classrooms, through the glass gallery of the Museum and over the temple's frieze and metopes – displayed in a fragmentary sequence with intermediate casts to recall for the missing pieces – constitute a terrain of powerful interactions for these children to improve contemporary appropriations of the ancient past as a reflection to its dismemberment.

The official Greek claim for the restitution of the Parthenon marbles to Athens from the British Museum, where they are exhibited after Lord Elgin had them violently removed from the monument in the early years of the 19th c. (St. Clair 1983), has already had a long history so far (Korka 2002; Hamilakis 2007: 243–286; Hitchens 2008). However, it has been particularly revived after the establishment of the new Acropolis Museum (Bernard Tschumi Architects

Fig. 1. Dominating view of the Parthenon and the Rock from the school's playground.

2009), emerging fast as another contemporary solidified landmark of western modernity. Further information on this history exceeds the scope of the present paper and can be easily researched in the vast literature and the worldwide web.

The project discussed here was launched in 2010 on the absolute initiative of the Acropolis School Parents Association, with the aim to contribute to the national claim by advocating an approach that would be focusing more on the privileged identities of the claimers as 'children' and 'neighbours' to the missing sculptures, rather than national citizens and generic heirs of a worldwide glorious cultural property. The project actually invited the Acropolis schoolchildren to narrate through visual artwork, their strong embodied and sensuous immediacy with the sculptures as their personal claim for the reconstitution of their dismembered neighbourhood. It also developed a plan to spread this claim to a worldwide audience with the publication of a multilingual anthology of the children's art. It is important to note that among the diverse formats the Greek campaign for the restitution of the Parthenon marbles had taken so far, this was the first time that the agenda involved children and locals, both sides engaged in the very same project. For those readers who are not acquainted with the degree the repatriation issue of the Parthenon sculptures is nationalised within the Greek context, in transcendence of any political and social divides, I should particularly emphasise that the involvement of children in this project is far from manipulating a politicised engagement or certain social grouping; for example, several non-Greek pupils of the school did equally share participatory enthusiasm and background ideals with children of Greek origin.

To be briefly informative on how the task developed, in the winter-spring of 2010 the Acropolis' schoolchildren were asked to join a series of art workshops held at the school beyond main class hours. The sessions were run under the guidance of instructors invited from the Museum of Greek Children's Art in Athens, a pioneering institution which is among a few of this kind worldwide, and is particularly specialized on teaching art history to young children, as well as on collecting and exhibiting children's art. The 82 kids from the school who responded to the invitation of the Parents Association were within the age range of 7 to 13. Before starting to make their own artworks, they had first followed a brief overview on the basic history of the Acropolis and the Parthenon, as well as the events concerning the removal of the sculptures. After this they were invited to compose their personal version of painted art, collage or combined technique within the general framework of the story about the marbles' presence, absence or return.

The project soon culminated and reached its partial fulfilment the same year, with the publication of the album anthology containing all artworks, under the title '*Our Sculptures Have their Homeland*' (KATSAROU-TZEVELEKI 2010), which was edited and published by the Parents Association with the raise of private funds. To further stress the primary axioms of the project, as local schoolchildren's privileged personal concern, authorship was ascribed to themselves under the definition of '*The young neighbours of the Acropolis*', featured above the title on the album's front cover (*Fig. 2*). The argument for their constantly experienced personal pain, due to the fragmented history, is however particularly strengthened by the powerful written appeal of the children – included in the form of a personal letter – cited before the artwork section to address their particular audience of worldwide friends of the same age. The letter was assembled from phrases and discourses spoken between the pupils during the painting sessions.

As the album from the beginning was planned to achieve the immediate attention of the international audience, this letter was repeated in 27 different languages, including most of the European, South-eastern and West Asian

Fig. 2. 'Our Sculptures Have their Homeland', *2010, book cover.*

languages, such as Chinese, Japanese, Korean, Indian, Arabic and Turkish. The strategic aim was to speak to the very heart of most of the people on the globe; all other complementary texts and captions of the book are in Greek, English and French. It is important to note that most of the translations were effectuated by the embassies of the relevant countries in Greece, some of them actually appointing the Greek debate for the Parthenon marbles with significant support from the side of the cultural heritage politics.

Publicity and further communication of the anthology has been an on-going aim of the project, in order to achieve a considerable contribution to the official campaign. The original artworks were displayed at a temporary exhibition at the Archaeological Museum of Delphi, Greece, during the spring of 2011. The album has so far been distributed to the press and to national and international agents, officials, academic scholars, and institutions, including the Swiss and the British local committees for the restitution of the Parthenon marbles; the latter actually welcomed the initiative with a one-page coverage in the winter 2012 issue of the campaign's online newsletter (KOUTSIKOU 2012). In Greece, the

album has been adopted, so far informally, by the state's current official advisory Committee established to promote the repatriation case. It has also received a very enthusiastic and touching response by Mr Spyros Mercouris, brother of the outstanding Greek myth personality Melina Mercouri who have passionately advocated the reunification issue as a Minister of Culture in the '80s, and whose giant pictures in front of the Parthenon are featuring as a commemoration to her and the nation's dream inside the 'Acropolis' Metro station in Athens.

Immediately below I will explore the dream narrative performances of the Acropolis schoolchildren as a foreground which has subconsciously revisited key conceptual trends on the constitutive role of the past in the ontology of the Hellenic nationhood. Later on I will attempt to situate these appropriations within the dense old on-going discussion on the evaluation of the antiquity's resourcefulness on a societal and epistemological level.

Enacted sculptures

The pupils' art has been materialised in six different thematic groups dealing with diverse versions of the appropriated presence or absence of the sculptures by the 'young neighbours'.

1.

In the first such group, the local kids and the antiquities appear as part of the same company of friends, the latter 'descending' from their temples to the pedestrian ways around Acropolis in order to play hide-and-seek and tag with the children; they are rocking on the swings at the school, are closed in the affectionate kids' embrace, they huddle up with the kids in a circle, carry them piggy back and are riding over the hills, or fly together with the children over the neighbourhood (*Fig. 3a, b*):

The young girl carrying a stool from the Panathenaic procession is looking at Adrianos who is running round and round. Stamatina is running amongst the Parthenon figures and the Caryatids who are looking at her. Sergios and Konstantinos are playing in the midst of the galloping horses from the Parthenon frieze. Areti is climbing on top of the Archaic colossal owl. The 'mourning Athena' and a horse-rider from the frieze are looking at Sotiris who is playing with his friends. Christina-Gavriela and goddess Athena from the pediment of Gigantomachy are playing at the slide and the swings at the school. Christos is firing a bow and arrow to goddess Athena who is hiding behind an olive tree, while the Caryatids are looking at him from further away. Spyros is lying on his back on a galloping horse that has sprung out of the Panathenaic procession of the Parthenon.

2.
In the second group, some of the personified sculptures are only basically sketched in pencil to denote that they live in exile. The convention – most likely suggested from the instructors, although some initiative from the part of the kids seems to have occurred in certain works – is visually inventive in terms of both style and symbolism: the grey and shadowy absences onto the contrastingly colourful hills and streets of the neighbourhood would depict the shape of the real sculptures rather than show a generic emptiness. The pertinence to the physicality, visibility and intransparency of the absent monuments in the imagery, should be exceptionally noted here as they signify constant sensuous perceptions in the children's memories. This surreal depiction is in a way alluding to the Platonic interpretation of visible things as ghosts or reflections of an ideal sphere (although probably the pupils did not consciously realise it as such), which here is meant to denote the ancient past and homeland cosmology, projecting memories onto the present life of the kids. More than that, the shadowy colour is additionally indicative of the children's sorrow towards these absences (*Fig. 4a, b*).

Vasiliki watches the shadow of a huge amphora that is missing from its stand at the park right outside the temple, and the grey sketch of a gigantic Pegasus that flies in front of a rainbow. Maria perceives the reflection of absent 'mourning goddess Athena' next to the temple and the shadow of a horse missing from the Parthenon frieze, galloping towards her. Chrysa, Sofia-Eleni, Stella and Aikaterina depict the reflections of the Caryatids and the Parthenon horses that are galloping under the hill. Zoi observes the reflections of the absent monuments on the pediment of the temple; Alexandra, Chara and Anastasia recognize the shadows of the absent columns and the cornice on the actual ancient ruins, while Orestis does see the reflection of an entire row of the missing Caryatids at its very original place at the temple of Erechtheion. Theodora, Margarita and Giasemi also depict the Caryatids, the 'mourning Athena', the colossal owl, and goddess Athena from the Gigantomachy as grey outline figures of absence still strolling around inside the colourful landscape.

Fig. 3a. Konstantinos Barkas, age 8, riding on the frieze's horse over the Acropolis.

3.

In the third group, transcending their physical limits, we watch the 'young neighbours' as they take the decisive initiative to bring these uprooted monuments back themselves. Contrary to the previous depictions of the absent sculptures as reflections, here the statues appear 'in person' – intentionally designated as a collage of photographs, so as to be illustrated realistically and vividly (*Fig. 5a, b*). Almost all images include the head of the horse from the Parthenon's east pediment:

Argyro and Dimitris position the Caryatids from Erechtheion and the chariot driver from the Parthenon frieze upright on top of train carriages. Natassa is

Fig. 3b. Areti Xenikaki, age 9, playing with the colossal owl and the Caryatids.

loading up the colossal owl on to a truck, Anastasia transports the enthroned gods Zeus and Hera from the Parthenon frieze and the horse from the east pediment by boat. Fanis however ties the same horse behind a wheeled vehicle together with the 'mourning Athena' and the young girl that carries the stool from the Panathenaic procession. Ioannis has loaded an enthroned god and the head of the horse from the east pediment on the same carriage, together with one Caryatid in upright position, while another Caryatid is hanging from a balloon. Artemis hangs a Caryatid from an airplane, while Athina suspends her same statue from another hot-air balloon together with the Parthenon's horse and the riders. Fotis has placed the horse in the hot-air balloon basket, but drags Athena from the Gigantomachy, wrapped in chains, lying down in the air. Panagiotis suspends

Fig. 4a. Zoi Domvrou-Ryan, age 7.

from the airplane one Caryatid as well as a slab from the Parthenon frieze and the statue of Athena Parthenos. Minoas has the Parthenon's man and the boy that unfold Athena's peplos on roller blades, and hangs the horse-riders from a hot-air balloon. Maria has put the enthroned and standing Parthenon frieze figures on top of train wagons. Finally Stefania manages to balance the horse from the

Fig. 4b. Maria Argyri, age 8, sketched pediment and statues denoting the absent antiquities.

pediment on a bicycle, and places the Caryatid upright and another head of a horse inside a van.

Fig. 5a. Fotis Kallivretakis, age 9, dragging by hands the Archaic statue of goddess Athena and hanging the Parthenon's east pediment horse from a hot-air balloon.

4.

We are not simply shown the sculptures moving in the artworks, but we are also capable to 'hear' their 'voices'. In the fourth group the canvas has become the field for a metaphysical dialogue taking place between the uprooted statues and those exhibited in Greece or the 'young neighbours'. The messages are mostly exchanged between the Caryatids expressing mutual pain for being forcefully separated, as well as their longing for return and the reassurance that they are now eagerly awaited by the children (*Fig. 6a, b*) (KOUTSIKOU 2012: Fig. 2). These dialogues directly refer to old local Athenian legends claiming that the spirit of the sculptures was upset from their violent dismembering, and that in the evenings the Caryatids left behind were heard crying for their lost 'sister' removed by Elgin (see references cited in HAMILAKIS 2007: 250).

Fig. 5b. Stefania Evangelisti, age 7, loading the horse on a bike and one Caryatid upright on a van.

'We want you to come back!!!' say the 'mourning Athena' and the colossal owl to the Caryatid in Anna-Maria's painting. Chara's owl is reassuring the Caryatid: 'Everyone is waiting for you!'. Dimitra imagines what the uprooted Caryatids are saying to their 'sisters' in Athens: 'We want to come back', and they are replying: 'We are waiting for you'. The Caryatids in Alexandros' painting wonder as they gaze upon the empty outline of the one who is missing: 'Where is our sister?'. Artemis proudly promises to the Caryatid: 'I will bring you to Greece so that everyone may admire you!'. Eleftheria is speaking vividly to the chariot-driver from the Panathenaic procession: 'We are looking forward to your return!'. Angelos reassures the horse from the Parthenon's east pediment: 'Everyone is waiting for you!'. And Chrysi has the galloping horse of the north frieze wondering: 'Where is my rider?'. In David's work the existing enthroned goddess Hera from the Parthenon frieze is speaking to her shadowy reflection: 'Come back, wherever you are'.

Fig. 6a. Dimitra Valadaki, age 11, (above) the sculptures saying: '- We want to come back'; *(below) the Caryatids replying:* '- We are waiting for you'.

Fig. 6b. Nikos Christopoulos, age 10, (the Caryatids saying) '- We want our sister!'

In this group of works, the repatriation of the sculptures is transformed into a metaphysical achievement from the part of the 'young neighbours' who are capable of bringing the ancient artefacts back upon their immediate interaction, excluding any state negotiations and permits. Nowhere does the imagery make any reference to museum space. It is only in two artworks that an indirect reference to the establishment of the Acropolis Museum is made; one such dialogue raises the old argument of the British side that the monuments, upon their restitution in Athens, would face severe danger in regards to preservation, confronted with its reversed update now posed by the establishment of the new Acropolis Museum:

Marizoi pleads with the Caryatid: 'Come back. Your place awaits you. Don't be afraid, you won't fall to ruin. Your sisters are waiting for you and crying. Come back as quickly as you can. Everyone awaits you'. Stella also shows the young girls that hold the stools in the north frieze of the Parthenon as being impatient: 'We want to return to our new home'.

The messages themselves are concise and minimalist and transfer power directly from the artwork to the soul of the viewer. The pinnacle is found in Eleni-Anna's work where the Caryatid, contrasted once more against a minimalist background, is dashing forward – as suggested by her leaning body – in order to take her place at the empty outline of her shape that awaits her (KOUTSIKOU 2012: Fig. 1):
'I'm already on my way!'

5.
In the fifth group, the sculptures are transported back to Athens through an imaginary, dream-like journey, imbued with the enthusiasm of friends that are about to join a long desired homeland. The journey is depicted as a transverse on the map of Europe, landing on the map of the neighbourhood around the Acropolis (Fig. 7):

For Margarita and Marvinia the Caryatid and the horse from the frieze reach the Acropolis and the Museum flying over Europe. Emmanouil depicts the statue of goddess Athena and the horse from the Parthenon pediment watching Europe from above on their return. For Anastasia Athena of the Gigantomachy is crossing Europe on her way back, while for Fay it is the 'mourning Athena' that journeys back on the map. Tzouliano imagines the entire row of the Caryatids crossing the map from the northern Europe to the Mediterranean. Christina-Foteini watches the front row of the Caryatids and the enthroned deities of the frieze landing in the neighbourhood. Evita and Giorgos envision the Caryatid and the colossal owl arriving mid air on the map of the Acropolis region. Alexandros also depicts on the same map two Caryatids, Eleonora six, Foivi shows the standing figures from the Parthenon east frieze and Elina a horse from the Panathenaic procession that comes back galloping.

6.
However, before they can take their final positions in their new home, either as old friends welcomed back by their companions or eagerly awaiting new neighbours, the sculptures first take a stroll in the neighbourhood, around the hills and the parks nearby the Sacred Rock.

Fig. 7: Fay Vlami, age 10, the Parthenon's east pediment horse and the 'mourning Athena' fly over Europe to Greece.

Michalis and Dimitris depict the enthroned gods of the Parthenon frieze already arrived under the hill of the Acropolis. Theodora's standing elders of the frieze have descended in the park of the Rock. Martha's young girls holding the stools in the Panathenaic procession, together with the seated gods Athena and Hephaestus are walking around the neighbourhood. Konstantinos imagines the small pony and the horse riders of the Parthenon frieze galloping on the pedestrian way, while Odysseas has put one slab from the frieze that has just arrived, straight outside the entrance of the new Acropolis Museum.

Transformed marbles

Transcending the physical boundaries of space and time, the 'young neighbours of the Acropolis' transformed the emblematic sculptures from the nearby monument into persons-protagonists of their contemporary life, where the ancient past is fully appropriated and assimilated. Their main companions are the Caryatids from the Erechtheion, horses, horse- and chariot-riders, enthroned gods and standing figures from the Parthenon frieze (HURWIT 2000; NEILS 2005). Featured are also the emblematic head of the horse from the Parthenon east pediment, together with the popular statue of goddess Athena from the pediment of the Gigantomachy originating from an Archaic temple on the Rock, the colossal marble owl of the same date, along with the beloved Classical relief of 'mourning goddess Athena' (Acrop. Museum, Athens, nos 681, 1347, 695). Some images of sculptures that do not originate from the hill are also adopted for their symbolic significance, such as the Roman statuette of Athena Parthenos (Nat. Arch. Museum, Athens, no 129), and a few other than sculptural antique elements, including Pegasus and ceramic amphorae.

The transformations were based on a particular iconographic style which the 'young neighbours' had been guided to create as an idiom of embodied imagery, where the ancient monuments were visually multiplied, enlarged or reduced, coloured and figured to look like dream images. They are situated in space like enacted beings, but without changing their physical sculptural form: e.g. even though the 'mourning Athena' and the Caryatids are supposed to be moving, they do not appear marching or with arms outstretched, do not turn their heads or have visible expressions on their faces. Instead they simply stand rigid and upright facing the viewer – apart from one exception where one Caryatid is laughing and another one is sitting on a swing. Similarly, Athena from the Gigantomachy, whatever she is depicted doing in the images, she is always closely adhering to the standing position of the real statue, which is with one arm outstretched and her robes hanging over it. The Archaic giant owl could have been portrayed as flying in the sky, yet it appears stiff and with its wings closed as is the ancient figure. Particularly in the case of the 'mourning Athena', the expression of sadness or thoughtfulness of her bent head, which contributes further to the dramatic personification of the figure in the painful story of the exiled sculptures, is standardised in the children's imagery. For the same reason, the figures from the Parthenon frieze are portrayed as separate individuals, or in pairs, in order to acquire an independent existence as persons in the children's narrations rather than be part of the ancient iconographic scenes.

A basic aspect of the surreal iconography of the 'young neighbours' is that it transcends the epistemological data that relate to Greece's request for the return of the Parthenon marbles, and that it restructures them in a new symbolic and metaphysic version. The choice of monuments enacted by the children does not appear to be determined by the official record of the monuments that have been detached from Acropolis to be exhibited abroad, neither does it correspond to the official list of monuments requested to be returned. It is absolutely expected that the 'young neighbours' freely shape their own version of absent sculptures in their artworks, perhaps choosing more among the most expressive figurative forms from the Acropolis (e.g. the 'mourning Athena') or the most popular (e.g. the Caryatids), rather than being concerned with following closely the accurate record of artefacts catalogued for reconstitution. In reality only one Caryatid is missing from Greece, the rest of them are in fact being exhibited at the Acropolis Museum, and not several as the children imagine in many of the their artworks, perhaps in their attempt to emphasise the gap left behind by the missing marbles. Moreover, the relief of the 'mourning Athena', the statue of Athena from the Gigantomachy and the colossal owl, which are here depicted as missing from Greece, are actually exhibited in the Acropolis Museum, while the Roman statuette of Athena Parthenos is also located at the National Archaeological Museum of Athens.

Therefore, in this iconographic narrative the emblematic figures of the past are reconstituted beyond any fastidious epistemological classifications and monolithic symbolisms to create a personal and participatory past for the 'young neighbours'; an eternal national grandeur that is appropriated through embodied action and contingency, and is absolutely lived in the *present*. But what is indeed the precise intrinsic essence of this memorialised base of Greek thought, which is being reconsidered and transformed here, and what are the avenues these background ideologies have travelled to reach the contemporary biographies of the Acropolis schoolchildren?

Emblematic past

Greek antiquity has repeatedly been discussed for the way it has come to represent the inherent resource of Hellenism, sometimes through a re-evaluation of its roots in the modernist paradigm, and often as a reproach for the lack of perspective with regards to the *future* of this ancient past. For the modern Greek nation, the common conclusion however remains, that the reception of this resourceful Classical heritage, either from a social or an epistemological aspect, has empowered a central ideology of a dream nationhood, imagined as 'secular cosmology' in Herzfeld's words (HERZFELD 1987: 28). According to him, this dream nation has to remain untouched, timeless and authentic, because it physically guarantees the exclusivity and longevity of Greek-ness (HERZFELD 1987; 2005). This is a theoretical construct which in many ways is not entirely indigenous, but is actually based on ideological frameworks situated within the contexts of western modernity. As such they have first transformed Greece's original Classical past to a myth of western ontology, and have then reintroduced it to Greece as a colonial hybrid. Under the pretence of the unique morphological exceptionality of Greek history and Classical antiquity, this ideological construct of awareness – incorporating trends of romanticism and idealism that were simultaneously being shaped in European thought – managed to generate Greece's core ideology and self-definition in modernity (HAMILAKIS – YALOURI 1996; KOTSAKIS 1998: 48; HAMILAKIS 2007; LOUKAKI 2008: 15). The extent to which this concept still remains powerful today is eloquently demonstrated in the country's political strategies, education, and its various official representations (PLANTZOS 2011; 2012a; 2012b; see also papers in BROWN – HAMILAKIS 2003 and in DAMASKOS – PLANTZOS 2008).

The symbol *par excellence* of this subjective grandeur is the Acropolis and particularly the Parthenon, which carry the weight of the present's promise to the Greek nation, to protect this unbroken existential relationship through time – by virtue of history and a sense of continuity – the glorious past, the present and the future of Greece. It is not surprising therefore, that the Acropolis, being an eternal stereotype at the top of the national stakes (YALOURI 2001: 49; 2010: 131; LOUKAKI 2008: 43), holds a prominent position that allows no negotiation and interaction with contemporary contexts. On the contrary, it is an emblem that seeks only enraptured observers and proud inheritors of a glorious past from within it, placed outside any dialogue that would risk it is exposed to any redefinition. This non-negotiable authority of the past that the Acropolis represents, is enforced by the official political agenda of the Greek state on its constant propagation and further reconfirmation of the illusionary national ontology: consequently the new Acropolis Museum has been inaugurated at the foot of the Sacred Rock in 2009 as an 'ark' of this national proof (PLANTZOS 2011), or the 'temple' of the nation (HAMILAKIS 2007), while the official national claim for the return of the Parthenon marbles from the British Museum follows the same theoretical line.

So it is under the weight of these undying, unilateral concepts that the Caryatids together with the most famous Parthenon's sculpted figures are invited to enter the field of the present by the 'young neighbours of the Acropolis'.

Embodied ideals

Despite this suffocating framework that surrounds the ancient past, as a solidified truth beyond any redefinition, the 'young neighbours' showed through their intervention that its rupture takes place from within and from the bottom to the top. As Hamilakis has put it: "The materiality of the nation is not only a state affair, but is simultaneously constructed from below and from above. The topos of the nation is constantly produced" (HAMILAKIS 2007: 17). Through the reframing of the traditional monumental emblems in their drawings, the pupils demonstrate that in parallel with this culturally appropriated landscape exists a world in which Classical antiquity can become, contrary to appearances, a point of common discourse between different poles within a structure of collective and individual negotiation. The Acropolis is indeed transformed from a universal and national stereotype to a dynamic terrain open to possible recontextualisations, which exist under or beyond its official status and participate in its palimpsest:

> "All members of the national body are potentially involved in the same way as the management of its ideological referents and principles is not only a constantly contested affair but it is open to all" (HAMILAKIS 2007: 18).

The 'young neighbours' have produced this rather multiple and cosmopolitan Acropolis out of the monologic ancient substance, imbuing it with subjective, dependent, questionable and indisputable properties.

The allegories of *space* and its *embodiment* within the schoolchildren's personal biographies and ethnographies, not naturalized in the sense of the real world, but rather qualified with a perspective of *idealism*, represent to my mind the main axes transcending the pupils' artworks.

The concept of space in the paintings should not simply be regarded as natural topographical depictions of the Acropolis and its environs, including open spaces, hills, swings, the Rock and the temples, the plants and the sky. It carries instead a strong social significance ascribed by the children's contemporary activity in the area, which is further added to the historical metaphors created for this locale by its ancient monumentality. This ethnographic aspect makes direct reference to a tangible, dynamic society of neighbours that consists of the families of the local residents living in the area south of the Acropolis under the domination of the Sacred Rock and the Museum, to which the particular social group of childhood friends and fellow pupils with the Primary School belong, having the pedestrian way and playground at the immediacy of the archaeological site as their main fields of activity.

Only to give a hint of the heavy load of glorious impetus that makes up the meaningful and powerful background of these children's ethnographies, I indicatively mention here as ordinary appropriations the key spots of the ancient topography, such as: the Roman Odeon (now Theatre) on the south slope of the Acropolis where Herodes Atticus held high art festivals for about 60 years, the early ancient theatre of Dionysus on the same side, the Athenian democracy's major landmarks at the hills of Pnyx hosting the ancient city's popular assemblies, and the Areios Pagos for the Court of Appeal. The newly constructed pedestrian zone, unifying the Acropolis and its environs with the Athenian Ancient Agora on the north slope and the Kerameikos cemetery further north west, as well as the Hadrian's Gate and the colossal temple of Zeus on the opposite side of the Rock, are high spot for locals and tourists expanding one's unmediated contact with the dominating landmarks. The bust of Melina Mercouri, at the start of the route to the Acropolis, and her commemorating giant pictures taken in front of the Parthenon that have been illustrating the 'Acropolis' Metro station, are another constant incentive for the 'young neighbours' to build renegotiable and reconstituting contemporary biographies of the ancient past. Finally, the new Acropolis Museum, recently established at the foot of the rock, is an outstanding example of modernity, essentially minimal and authoritative, heavily achromous and dogmatic. Although, in a way it can be seen as attempting to assimilate the transformative processes of the 20th c. to achieve some immediacy with the viewer.

The group of children frequenting these emblematic landmarks is socially diverse, as it encompasses several different social groups, for example, from different nationalities or financial standing. The group is not exclusionary, at least at the level of its self-definition, which is to form friendships and play together in the neighbourhood, a fusion that is further demonstrated in the inclusion of many 'young neighbours' of non-Greek nationality in the local-managed artwork project. In fact the glory of the dream nation has been very appealing for several immigrants to share Hellenism and thus update their cultural status within the Greek society.

The immersion of the Acropolis within a multitude of social groupings holds a prominent place in the anthropological discussion of the monument (YALOURI 2001: 101; LOUKAKI 2008: 15), however they mostly reference a national, urban or other special level such as, for example, the tourists visiting the site. In fact the social framing of the monument within the community of local kids, and particularly playfellows who attribute to the emblematic site their very local and very personal understandings, is for the first time recorded in the collection of these artworks by the 'young neighbours of the Acropolis' discussed here. In this particular visual ethnography, the illustrated monuments exist as embedded with the schoolchildren in the locale, and as absolutely sensuously embodied by the kids, so as to constitute a desired *homeland* that would constantly verify the enduring continuity of the ancestry of Hellenism. Such a homeland has been discussed by GOURGOURIS (1996: 42) and HAMILAKIS (2007: 16), who has reassigned the Foucauldian metaphor of *heterotopia* to signify the imaginary topos that expands beyond any spatial dimension of the ancient land, to include all the possible figures and symbolic materialisations that would denote the ideal values of Hellenism. The exiled Parthenon sculptures potentially conceptualise such a mythical heterotopic homeland of the nation, invested with the desire for return.

But in the neighbours' artworks it is not only such a generic national framework that is recalled, but, more than that, it is the particular ethnography of this very local neighbourhood and its most privileged connection to the embodied cultural property that employ the imagery of the sculptures as the heterotopic homeland. It is the local ideal of the ancestral space that is narrated here, generated from the embodied experience with its heterotopias (HAMILAKIS 2007: 18), achieved through the bodily senses and emotional contact in the course of the ordinary daily practices and routines of the children.

Embodiment is a key concept of this project. It transfers the longue durée, based on an individual experiential view of randomness and contingency, to interaction on the small scale, responding to contemporary disciplinary trends of cultural anthropology (HAMILAKIS 2002). It was in fact already in the 19th century that empiricism and romantic neo-humanism improved the antiquity's corporeal dimensions within the dominant ideological construct of German classicism (MARCHAND 1996: 7), by reprocessing older experiences of embodied appropriations of the past, deriving from long before modernity (HAMILAKIS 2011). The Acropolis Rock and the Parthenon are loved, as well as admired, awed or even feared, and invested with metaphysical powers in the local legends (PHILIPPIDES 1994). They are dreamt as ideals and morals (GOURGOURIS 1996: 29), sanitized and deified (ALONSO 1988; LEONTIS 1995; HAMILAKIS – YALOURI 1996; 1999; YALOURI 2001: 137), as clearly shown in the expression 'Sacred Rock' for the hill of the Acropolis.

The Acropolis pupils' imagery draws direct references from older stereotyped idyllic images of this 'pilgrimage' to the Rock and its environs painted by the 16th and 19th century travellers. In these paintings the shepherds, who herd their stock near the Acropolis, are depicted as an organic part of the ancient monumental landscape (MARCHAND 1996: 24; GOURGOURIS 1996: 196), and are by extension dreamed to be privileged people in history, thereby generating an enduring legacy in the imagination of western European thought, reaching all the way up to these 'young neighbours of the Acropolis' of today.

It is this same perception on the dreamed past, that subconsciously transfers powerful sensuous and corporeal qualities through its constant experience by mere breath and vision, that the famous storyteller Hans Christian Andersen also noted in regard to the children he saw wandering around the ruins of the Acropolis during his own pilgrimage to the hill in 1841:

> *"How enviably fortunate Greek children are to play, albeit blissfully unaware, amongst the classical places, breathing ancient dust in the shadow of the Acropolis* [...]*"* (ANDERSEN 1988: 54).

Those early 'young neighbours' of the Acropolis are already regarded to carry the ontology of the Greek nation on their shoulders at that time, though 'unaware' of its emblematic significance within the European context. On the other end of the line contemporary children – their antecedents in the area – probably make similar appropriations but have been driven to become conscious of their heritage, with particular awareness of their very specific position within their national and local history, and develop interactive discourses with it. The former are depicted in the iconographies of others and are thus externally signified, the latter constitute themselves in their own individual significations as deeply existential.

Trauma in light and marble

The 'young neighbours' do not passively and automatically state some national identity and great history of the country, but propagate the ancient sculptures through a powerfully transformative and interchangeable bodily command of the monuments.

This corporeal vision endow the topographical features of the Acropolis neighbourhood with the deterministic and idealised self-explanatory symbolisms of grandiose descent and moral lessons, which were otherwise ascribed to the national terrain: the *marble* of divine representation, eternal duration and reflection of beauty; the *light* of the clarity and radiance of civilisation, and the *ruins* of the traumatic history and the nostalgia for the antique glory.

Here are the words of the 'young neighbours', addressing the children all over the world with the awareness of constituting a living heir of antiquity by determinism:

> *"Dear friends all over the world,*
> *We are pupils of the 70th Athens Primary School which is located near the Acropolis of Athens. Out of our class windows and from the courtyard of our school we have a view of the Acropolis, the Parthenon and the statues that are housed in the New Acropolis Museum next to us.*
> *We feel we can talk to them. We know them by their names. They talk to us too. They talk in our hearts and thoughts. However some are missing. In their places there are replicas made of stucco. These are lifeless, they are strangers to us. They do not listen to us nor do they talk to us. The original ones are somewhere else, far away from us. We ask that they come back to their rightful place near us. That they come back home across our school.*
> *That they come back to the Light that created them."*
> (KATSAROU-TZEVELEKI 2010: 15)

The 'young neighbours' personal trauma, owing to the ruined antiquity, appears as a most powerful partaking of the embodied past, sending to the world the most forceful appeal for the reunification of their dismembered cultural body. The intimacy between schoolchildren and sculptures is the advantage of this appeal in comparison to the generic national claim of the Greek state. Its privilege resides in the existential bond between the two and the uniting force with their homeland which is a physical organism and experiences the dramatic dislocation of its members from the body of its history.

These life episodes are ethnographies of shared moments of play, but also of protests and cries for the disembodied community of neighbours. However they also extend to stories of powerful activism from the part of the children, stressing through the immediacy of communication their determination to materialize the repatriation of the sculptures only via the mutual consent of the children and the sculpted figures, beyond the scope of any external intervention. Their close intimacy renders their claim just as powerful as any official negotiation on the issue.

In Greece this sensuous and individual corporeal understanding of antiquity that the Acropolis schoolchildren demonstrate, has been occurring in personal contexts aside the institutionalised frameworks of nationalist ideals and in an unbroken continuity with 'pre-modern' ideologies (HAMILAKIS 2011), sometimes suffocating within the big dogmas, but always becoming constitutive of the past, even if negative, critical or dismissive of it. This particular Greek experience has generated an idiom of modernity that is specifically historical and has been extensively inspiring for later Greek ideologies, including the project in discussion.

Critical regionalism

The synthesis of Hellenism from the fusion of classical authority with contingent embodied and indigenous qualities, has been the product of a long fermenting vision that goes back to the so-called 'Generation of the 30s' in the first half of the 20th c. (see papers in TZIOVAS 1997; 2001). This inspired intellectual initiative, negotiated the aspect of Greek-ness 'between the local and the global' (YALOURI 2001; 2010; PLANTZOS 2012a: 229) by challenging it to an esoteric quest for its indigenous heritage properties to which it attributed even further universality. This project constituted a mythical, exotic and exportable Greek-ness out of the national paradigm, one that aimed at recolonizing western thought. This is something which still had to be fulfilled since the particular historical circumstances of the first half of the 20th century, demanded the immediate self-affirmation of the Greek state, and inspired the nation's ideology across all social, cultural and political legitimacies. A significant agenda within this project was to merge the sensuous and corporeal material of the Greek experience, in a post-romanticism and post-idealism context, to re-empower the grand narrative of the nation.

As further discussion is beyond the scope of the present paper, I will only refer to what is of immediate interest for us here, namely two certain architectural projects that were then run in the area of Acropolis to enhance the power of the nation's ideology in alignment with the concept. Both are still visible in the contemporary landscape where the 'young neighbours' live and play, so that, apart from their acquaintance with the nation's ideals through the official heritage strategies and generic ideological background, the Acropolis schoolchildren constantly experience by immediate corporeal connection the heritage of this influential trend where antiquity is recreated and recontextualised.

In the 1950s the architect Dimitris Pikionis responded to the project of the landscaping work in the Precinct of the Acropolis, by developing a language of architectural regionalism (PIKIONI – PIKIONI 2010; BASTEA 2003), seeking its drastic incentives in the ontological reference of this particular landscape to the past. By utilising rough-finished marble paving slabs in various shapes that appear irregular, yet are strictly geometric, Pikionis envisioned classical architectural features inside the contemporary and inherited present. Thus, in the world of the 'young neighbours' exists a terrain of a recreated past.

Another influential architect of this generation, Patroklos Karantinos (GIAKOUMAKATOS 2003) had designed and constructed in 1931 the building of the Primary School in Kallisperi street, opposite the south slope of the Acropolis, this very same building where the 'young neighbours' are still housed today. Equally inspired by the powerful pursuit to capture the contemporaneity of the past and its retransformation, rather than any mimetic reproduction of antiquity, he disputed the neoclassical trend and instead effectuated a minimal geometric building of two floors which faced the hill on its long side and was permeable to light and the unmediated view of the Parthenon through big and long windows on the upper floor. Some years later, the same architect ran functional modifications to the old Museum of Acropolis in line with the same ideological perspective.

The influential current of the 1930s further solidified that Greek-ness is rooted in the ancient soil, and updated the concept of the attachment to the earth and its appropriation as major cults for the nation's reaffirmation as a dream culture (HAMILAKIS – YALOURI 1999). In the words of Loukaki:

> "Evidence from the press and literature indicates that the adoration of the Parthenon and the plethora of ways in which it is embedded into Greek society are remarkable, involving a combination of highly intermingled scholarly and popular elements" (LOUKAKI 2008: 253).

The official strategies propagated by the Greek state solidified this sanctification of the past (HAMILAKIS 2003: 56). The national project, deploying the methodology of national ceremonies and bio-politics (FOUCAULT 1990: 135–159), is still thriving today. The revival of Greek classical drama in the 1950s, with the aim to establish cultural continuity between modern and ancient Greece, is such an example of the use of a technology of embodiment, as are the parading statues that have come to life in the Opening Ceremony of the Athens Olympic Games in 2004 (PLANTZOS 2008: 13; 2012b: 228). Plantzos writes that these

> "performative bodies are deployed through the pageants and the staged re-enactments as a message abroad, certainly, but mostly as a pedagogic and disciplinary device at home, as a lesson of what to do with your (national) self. As a result, contemporary Greece deploys a self-made neoclassical bio-politics in order to construct itself as a modern European nation, Hellenist as well as Hellenic" (PLANTZOS 2012b: 166).

In respect to the young age, museum activities and didactics directed to children in Greece has considerably emphasised the deployment of such bodily and personification pedagogies to perpetuate and solidify the appropriation of the past. However, the resourceful background of the recontextualisation that was inherited from the generation of the 1930s is usually reproduced in simplistic ways that achieve the intellectual subjugation of schoolchildren to a given national dogma, but do not cultivate any further reconsideration.

This aspect should be particularly emphasised here for our better understanding of the ideological background behind the Parents Association-managed project. The 'young neighbours' carry multicontextual ideologies varying from the nations grand narratives of the classical past, to their very individual appropriations of the antiquities in their everyday life in their immediate locale. The engagement of some children of non-Greek origin with the project should be also noted as part of this contextual diversity. The collaboration with the Museum of Greek

Children's Art has been a key choice for the project and must have been chosen for its seminal research approach on art pedagogy and its deep insights on the history of Hellenic culture that have been creative for conventional Greek modernist ideologies, however without exploring extreme reconsiderations. For this reason its involvement has in many ways secured a respectful revisit of the nation's ideals and an integration and incorporation of any possible diversified approaches, for example from the pupils of non-Greek origin, and their alignment with the narrative of the admired nation. I have already noted above that sharing the glory of Hellenism has indeed been very much desired by some of them in order to upgrade their status within the Greek society.

Indigenous claim

The personification of the ancient monuments and their enlivening as participants in a dialogue has connected the receiver of history, here each child, with the distant past. The past is transported to the present and becomes lived. In this manner, the past is familiar, the child's own participatory field, and a terrain which he or she is shaping. This kind of discourse has generated the transformation of the ideological axis of the paintings from the national, grand narrative to the local, an important step to take towards the empowerment of the more personal aspects of culture. Thus, aside from being an 'Identity Strengthener' and an 'Eternal Truthteller' in the earlier words of Loukaki, the Acropolis also becomes for the kids what she would have called a 'Personal Thriller',

> "conferring not only a sense of individual identity, but also acting as a source of inspiration, of personal representational capacities, like the experience of time and space and the motivation of individual imagination" (Loukaki 2008: 17).

However this significant step towards the recontextualisation of the monolithic evolutionist past, by emphasising the embodied experience of the pupils with their cultural heritage, has essentially utilised the material available from the axiomatic stereotyped ideals of the national and the global scale, easily identifiable with a perpetual and mythical Hellenism. By drawing resources from earlier cultural transformations of the palimpsest of antiquity and its 'authentic' glorious background back to the beginning, the 'young neighbours' have built a singular idiom of their local homeland, a local fantasy as a kind of critical regionalism or what Hamilakis has called 'indigenous Hellenism' (HAMILAKIS 2007: 27).

Within this framework, the 'young neighbours' have probably not managed to surpass the dominant stereotypes of eternal truth to any large extent, and possibly failed to raise an unconventional critique that would be open to silenced and sceptical sides of the past; a language that would interrogate singular authoritative explanations and conservative rhetoric and would explore contemporary plurality and diversity, not being afraid of upcoming clashes. Indeed, HAMILAKIS (2012) forcefully emphasized that it is on this ground the confrontation of localism with vernacular discourses will empower societies in the future. What he called the more profound and multi-faceted clash between the logic of the capital and the logic of community, or the clash between uni-temporal past and multiple archaeologies enacting multiple, simultaneous times, has not been achieved by the children's project which decided to stay more connected to the fundamental and primordial ideals of Greek-ness rather than lean to the side of a more pronounced indigenous voice.

I have to admit that it was with disappointment that I realised that the response to the schoolchildren's project from the various audiences such as in the museum, the institutions, scholars or other sites where it has been presented, was attached to the universality of the Parthenon marbles rhetoric and to a colonial and national essentialism. Although enthusiastic for the embodied and emotional art, and touched by that, responses have so far mostly failed in captivating the indigenous element of the kids' narratives. External viewers of the artworks or readers of the anthology are always caught by the national aura of the trauma and the fact that it is expressed by children, the personal aspect of appropriation still evading together with the social significance of the local initiative on the community's cultural patrimony. Similarly this concept on the local claim for the homeland advocated by the indigenous community of the schoolchildren was not particularly realized either by the national and international visitors of the temporary exhibition of the pupils' artwork held at the Archaeological Museum of Delphi in 2011, where a book of notes was specifically set to assemble individual reactions. Their response primarily engaged with the clichés, debating the Acropolis marbles as colonial cultural heritage and world shared monument in the British Museum:

> "All of your rocks belong to us" (signed by 'London').
> "I think they need to get over it because we have them and we will destroy you if you try to get them back."
> "Do the children understand what would have happened to the statues had they not been taken?"

If it did not go for mocking:

> "I think that we have enough Goddamn rocks here anyway, let them keep the statues, they are ugly anyway!"

And it was only one certain comment, possibly by a child, to notice the participatory role of the kids in the artworks:

> "... I like that some kids put also themselves in the picture!"

And yet the children's art narrations have, achieved a refreshing reframing of modernity, not solely because of the inventive imagery, the strong emotional qualities and sensuous excitement of the project, or the activism for the heritage politics, but primarily thanks to this very empowerment of the social involvement of their indigenousness as an active social claim, which they developed without social discriminations and exclusions, as each of the local members could participate on equal terms without being comparatively evaluated.

The effectiveness of the 'young neighbours' idiom of social engagement on the support of a local social claim, remains to be evaluated through time, in potentially many subsequent events of renegotiation. The symbolic imagery of the schoolchildren have constituted polysemous biographies that await their further redefinition within the contemporary and future social contexts, as we have seen it happening with other renegotiations of the cultural heritage. Similar to the Acropolis project, such recontextualisations of antiquity often utilise its more dogmatic features to achieve a broad range of social negotiations from conventional reproduction by official agents, to the extreme end of a categorical or even defiant intervention within non-institutionalised milieus. The representations of kouroi, for example, from a political, anarchist or sexual disposition (PLANTZOS 2012a: 227), the use of the Acropolis for the raising of giant banners along the ancient fortification wall with messages from the trade unions of the state employees of the archaeological site, or by political parties for dominant national or global protests, relocate the past from the timelessness and austerity and embed it inside a concrete and timely social context. A similar particular social and historical participation for their claim was wishfully evoked by the 'young neighbours' and their parents who invited this personal, experiential and non-exclusionary discourse of the emblematic Acropolis and the Parthenon and the transformation of the subsequent conventions produced all the way across the palimpsest from the antiquity to modernity.

Therefore I want to particularly stress here as a conclusion that the local social community in the environs of the Acropolis so far existed and reacted on the side of the national agenda and in general on the side of any cultural heritage strategy and touristic politics for the monument par excellence, mostly resting in observing or admiring its land, rather than expressing any reaction to its powerful resourceful classical property. The local-managed project of the Acropolis parents was in fact the first initiative ever undertaken by the indigenous people of this glorious place to speak for their own past – including any past and not only the major restitution issue. It was the voice that they themselves cried to the world, the language they wrote, and the story they narrated, on their own choice, not on any official or state support. However conventional the project turned to be in respect of the national ideals, its launching on the very local decision is its very great moment of significance.

As a final remark I will go back to the point made in the beginning of this very same paper, where I advocated that the local school's Parents Association project is beyond any politicised engagement. I now want to add that, although it not an issue manipulated by a certain political area, yet, it does involve a critical degree of political and ethical awareness in the sense of actions taken on a local initiative to advocate a stance of a 'home' self and of a local social consensus. It is precisely thanks to this multi-temporal and multi-sited contextualisation of life stories that this certain patrimony, as any past, would acquire meaningfulness, even if it encompasses a hybridization and deconstruction of the dogmas. I would end this paper with what I consider to be the broader lesson taken from this, that in remediating the Acropolis as a-temporal and fixed, resurrected from the past, these personal narratives discussed here, fluid in time, emphasise its timeliness and transcendence of the past-present from the social geography of the micro-scale. The momentary interactions and narrations, realist as much as idealist – made of the stuff of national symbols but with a personal mediation – emerge a vision of overcoming stalemate regimes of culture and a great potential for local social ideologies for the past to come into sight.

Acknowledgements

I am grateful to the editors of this volume for inviting me to contribute and for their stimulating comments that have greatly improved my paper. My involvement in this project is twofold, as the mother of a child who took part in the project, but also through my contribution to the copy editing of the publication of the anthology. I would like to thank the 2008-10 Board of the Parents Association of the 70th Primary School of Athens for inviting me to collaborate, and the 2011-13 Board for giving me permission to reprint the images from the album collection. My special thanks should go to Athanassia Psalti, Director of the Archaeological Museum at Delphi, who had the idea of organising the temporary exhibition of the artworks of the 'young neighbours' at the Archaeological Museum of Delphi (March 15 – May 1st, 2011). My special appreciation goes to museologist Nagia Dalakoura for her collaboration on the same project. I am grateful to Dr Maria Relaki who has translated an early draft of my text to English.

References

ALONSO, ANN MARIA (1988), The effects of truth: re-presentations of the past and imagining of continuity. *Journal of Historical Sociology* 1.1, 33–57.

ANDERSEN, HANS CHRISTIAN (1988), *A poet's bazaar: a journey to Greece, Turkey and up the Danube*. Translated by Grace Thornton. New York.

BASTEA, ELENA (2003), Dimitris Pikionis and Sedad Eldem: Parallel reflections of vernacular and national architecture. In: Keith S. Brown – Yannis Hamilakis (eds), *The usable past: Greek metahistories*. Lanham – Oxford. 147–170.

BERNARD TSCHUMI ARCHITECTS (ed.) (2009), *The new Acropolis Museum*. New York.

BROWN, KEITH S. – YANNIS HAMILAKIS (eds) (2003), *The usable past: Greek metahistories*. Lanham – Oxford.

DAMASKOS, DIMITRIS – PLANTZOS DIMITRIS (eds) (2008), *A singular antiquity: archaeology and Hellenic identity in 20th-century Greece*. Athens.

FOUCAULT, MICHEL (1990), *The history of sexuality. An introduction*. New York.

GIAKOUMAKATOS, ANDREAS (2003), *Στοιχεία για τη νεότερη ελληνική αρχιτεκτονική: Πάτροκλος Καραντινός [Elements on contemporary Greek architecture: Patroklos Karantinos]*. Athens.

GOURGOURIS, STATHIS (1996), *Dream nation: Enlightment, consolidation and the institution of modern Greece*. Stanford.

HAMILAKIS, YANNIS (2002), Experience and corporeality. Introduction. In: Yannis Hamilakis – Mark Pluciennik – Sarah Tarlow (eds), *Thinking through the body. Archaeologies of corporeality*. New York. 99–103.

HAMILAKIS, YANNIS (2003), Learn history! Antiquity, national narrative and history in Greek educational textbooks. In: Keith S. Brown – Yannis Hamilakis (eds), *The usable past: Greek metahistories.* Lanham – Oxford. 39–67.

HAMILAKIS, YANNIS (2007), *The nation and its ruins. Antiquity, archaeology and national imagination in Greece*. Oxford.

HAMILAKIS, YANNIS (2011), Indigenous archaeologies in Ottoman Greece. In: Zainab Bahrani – Zeynep Celik – Edhem Eldem (eds), *Scramble for the past: the story of archaeology in the Ottoman Empire, 1733-1914*. Istanbul. 49–69.

HAMILAKIS, YANNIS (2012), Are we postcolonial yet? Tales from the battlefield. *Archaeologies: The Journal of the World Archaeological Congress* 8.1, 67–76.

HAMILAKIS, YANNIS – ELEANA YALOURI (1996), Antiquities as symbolic capital in modern Greece. *Antiquity* 70, 117–129.

HAMILAKIS, YANNIS – ELEANA YALOURI (1999), Sacralising the past: Cults of archaeology in modern Greece. *Archaeological Dialogues* 6.2, 115–159.

HERZFELD, MICHAEL (1987), *Anthropology through the looking-glass: Critical ethnography in the margins of Europe*. Cambridge.

HERZFELD, MICHAEL (2005), *Cultural intimacy: Social poetics in the nation-state*. New York.

HITCHENS, CHRISTOPHER (2008), *The Parthenon Marbles: The case for reunification*. London – New York.

HURWIT, JEFFREY M. (2000), *The Athenian Acropolis: history, mythology, and archaeology from the Neolithic era to the present*. Cambridge.

KATSAROU-TZEVELEKI, STELLA (ed.) (2010), *The young neighbors of the Acropolis: Our sculptures have their homeland ...* Trilingual publication. Parents' Association of the 70th Primary School at Athens in collaboration with the Museum of Greek Children's Art. Athens.

KORKA, ELENA (2002), *The reunification of the Parthenon Marbles*. Athens.

KOTSAKIS, KOSTAS (1998), The past is ours. Images of Greek Macedonia. In: Lynn Meskell (ed.), *Archaeology under fire*. London – New York. 44–67.

KOUTSIKOU, MARIA (2012), The young neighbours of the Acropolis make a visual appeal, *Newsletter of the Marbles Reunited Campaign* 9, 3. (http://www.marblesreunited.org.uk/news/newsletters [7.7.2013]).

LEONTIS, ARTEMIS (1995), *Topographies of Hellenism. Mapping the homeland*. Ithaca.

LOUKAKI, ARGYRO (2008), *Living ruins, value conflicts*. Aldershot.

MARCHAND, SUZANNE L. (1996), *Down from Olympus. Archaeology and philhellenism in Germany 1975-1970*. Princeton.

NEILS, JENIFER (2005), *The Parthenon: From antiquity to the present*. Cambridge.

PHILIPPIDES, DIMITRIS (1994), The Parthenon as appreciated by Greek society. In: Panayotis Tournikiotis (ed.), *The Parthenon and its impact in modern times*. Athens. 278–308.

PIKIONI, AGNI – DORA ROKOU-PIKIONI (eds) (2010), *Δημήτρης Πικιώνης 1887-1968 [Dimitris Pikionis 1887-1968]*. Athens.

PLANTZOS, DIMITRIS (2008), Archaeology and Hellenic identity, 1896–2004: the frustrated vision. In: Dimitris Damaskos – Dimitris Plantzos (eds), *A singular antiquity: archaeology and Hellenic identity in twentieth-century Greece*. Athens. 9–32.

PLANTZOS, DIMITRIS (2011), 'Il n'y a pas de hors-texte': Το Μουσείο της Ακρόπολης και τα απόνερα του ιδεοαλισμού [The Acropolis Museum and the wake of idealism]. *Tetradia Mouseiologias* 7, 23–29.

PLANTZOS, DIMITRIS (2012a), The *kouros* of Keratea: constructing subaltern pasts in contemporary Greece. *Journal of Social Archaeology* 12.2, 220–244.

PLANTZOS, DIMITRIS (2012b), The glory that was not: embodying the classical in contemporary Greece. *Interactions: Studies in Communication & Culture* 3.2, 147–171.

ST. CLAIR, WILLIAM (1983), *Lord Elgin and the Marbles*. Oxford – New York.

TZIOVAS, DIMITRIS (ed.) (1997), *Greek modernism and beyond: essays in honor of Peter Bien*. Lanham.

TZIOVAS, DIMITRIS (2001), Beyond the Acropolis: rethinking neohellenism. *Journal of Modern Greek Studies* 19, 189–220.

YALOURI, ELEANA (2001), *The Acropolis: Global fame, local claim*. Oxford.

YALOURI, ELEANA (2010), Between the local and the global: the Athenian Acropolis as both national and world monument. In: Anna Stroulia – Susan Buck Sutton (eds), *Archaeology in situ: sites, archaeology, and communities in Greece.* Greek Studies: Interdisciplinary Approaches. Lanham – Oxford. 131–158.

Public versus professional perceptions of an invisible heritage: A Greek case study

JOHN BINTLIFF

Abstract:
Greece is a country where particular phases of the Past and events in Athens and at other famous sites (panhellenic sanctuaries, Bronze Age palaces) dominate popular knowledge and wider tourism. The vast majority of the Greek landscape lacks local or international points of heritage presentation and hence its occupants are disempowered from their own immediate history. This paper will discuss how nonetheless local communities have tried to create their own narratives and how professional archaeologists can assist them into a more informed engagement with the rich monuments to be found in all Greek landscapes.

In the summer of 2012 my co-director Anthony Snodgrass and I were invited to give public lectures to village history societies, by leading figures of two of the communities whose nearest ancient town and countryside we have been researching into for some decades, Askra and Koroneia. Their enthusiasm to discover what our project had found out regarding their local history was inspiring. It reminded us that such villages, and especially their local professionals and retired professionals returned to their origin-communities, were responsible for periodicals regarding traditional life, pamphlets and even monographs on local history, archaeology and folklore. Yet in discussions with villagers and within such publications one can often find quite a mixture of local myth and folk-tales, memories from the oldest people in the community, and quotations obtained by research into academic output. Usually there is a common stock of rather legendary concepts of local history known to most inhabitants, then amongst the small group of the self-identifying intelligentsia (schoolteachers, doctors, architects, former generals and admirals and even university lecturers) one finds also knowledge of published research and of local studies of dialect, folklore, historic photographs, paintings and oral history.

Given that History in Greece has till very recently been taught in schools and universities from a national viewpoint, emphasizing Athens and the Classical past, with little or no scope for Local History, most rural communities find themselves excluded from established History, and their communities relegated to the margins of events. In the widespread absence till recently of a lively intelligentsia within them, members of rural communities have usually been led to create their own local traditions, in order to give themselves a sense of 'a place in history' (cf. FORBES 2009; this volume; DELTSOU 2009). This paper will focus on these alternative stories about communal pasts.

John Barrett, in his book *Fragments from Antiquity* (BARRETT 1994) has argued for a simple explanation of how the time of the individual and the event can be linked to longer-term historical processes – through the role of memory. The handing on of memories of people, places, and things – either verbally, through texts, or through monumentalization – is argued by Barrett to create the means through which, following the ideas of the sociologist Anthony Giddens, individual human agents can tie their beliefs and actions (what Giddens calls the daily acts of reproduction of society), into a potentially infinite future time. However I have serious difficulties with the hypothesis that individual or collective memory – oral, textual, or monumental – systematically conveys the time of events into the medium and long term, without being transformed in the process, through memory loss, selective survival, and reinterpretation. We are all aware today of the phenomenon of 'false memory', and also of the documented evidence that individuals present at the same event have diverse recollections of what occurred. We now also have excellent studies of the ways in which historic and legendary traditions suffer progressive editing and transformation across the generations when oral memory is central to their transmission (HALL 1997).

In connection with the idea of the artificial creation of an historical memory, the following example is of interest. If modern communities have access to academic texts then it is indeed possible to create a 'new memory' of local pasts previously unknown to their recent inhabitants. The village of Erimokastro in Central Greece, for example, was called after a ruined Late Roman fort beside it, and was only renamed, after the adjacent ancient city of Thespiai, as 'Thespies' during the process of Hellenization of placenames which occurred from the end of the 19th and through the early 20th centuries AD (ALEXANDRI 2002). The choice of name was possible from the work of foreign archaeologists and historians who had identified the ancient site in the plain below the village from coins and inscriptions during the 19th century. Since 'rediscovering' its deep past, as a community possibly descended from an ancient Greek city, attempts have been made to enhance the village's otherwise minimal history. The latest is a striking monument erected on the village entrance by a leading wealthy citizen (*Fig. 1*). Deliberately recalling the appearance of the 300 Spartans who sacrificed themselves at Thermopylae and which gave rise to a far-from-historically-accurate Hollywood blockbuster movie, our Greek warrior in fact represents the 700 soldiers of ancient Thespiai whom (we forgot!) died alongside the Spartans. Worth mentioning – the face is modelled on the businessman who paid for the monument.

Fig. 1.

The text is a remarkable creation, and also has its unintentionally humorous moments.

300 + 700. THE LAST DEFENDERS OF THERMOPYLAE

In August 480 from this very place the 700 Thespians voluntarily set off for Thermopylae under General Demophilus and the commanders Xenokratides, Dithyramus and Protokreon. Following the Efialtes treason, the battle was lost. The Spartans who were forced to stay, ordered the allies to leave the battle. You see, Leonidas, King of Sparta had the general command. The order was consented to by many, but not the Thespians. This was not agreeable to Leonidas. You, Leonidas, are not taking our glory away, replied Dithyramus, so they stayed there and sacrificed themselves ...

They did not desert Leonidas and his peers but they stayed there and died together. And they were led by Demophilus the Diadromeus [Herodotus].

Still verbal history continues for 25 centuries to present the Spartans only. This is offensive for me and I feel the duty to castigate this injustice. It's as if Thermopylae was only a Spartan issue. You foreigner, while you read these lines, do not forget, glory belongs to all those who gave their life for the country.

Constantinos D. Liakos-Kelepis. Thespian.

My second example, also from Central Greece, concerns an isolated ruined monument which lies in the lands farmed by the modern villagers of Agios Georgios (*Fig. 2a–b*). It is in fact part of a Medieval feudal tower erected by French Crusader knights who overran the region in the 13th century AD. But it also lies at the foot of a large hill where lay the ancient city of Koroneia. The villagers were alerted to the existence of the ancient city by academics from the later 19th century, but remain unaware of the plan or chronology of fragmentary ruins found scattered over the site.

In 2009, at the entrance to the modern village, the Mayor had a monument erected (*Fig. 3*). It is in fact a passable reproduction of the Crusader tower, but since its date and role are unknown to the community, and it is the largest visible ruin on the ancient city, it has been assumed to be a part of the ancient Greek city wall – and hence worthy to mark the village entry. Although the community will be disappointed to learn that they are instead celebrating Western conquerors, nonetheless it marks an impressive local attempt to anchor their otherwise hardly documented village into a famous past.

My third example derives from nearby Albania, a recent excavation of a Late Bronze to Early Iron Age tumulus burial mound at Lofkend (PAPADOPOULOS *et al.* 2008).

> "In the context of Lofkend, it is interesting to note what the local inhabitants of the area thought of the mound prior to excavation. When asked what the tumulus represents, most if not all of the local villagers of nearby Ngrancija and Gjinoqara considered the mound as a burial place not for the long dead, but for those just beyond the grasp of memory. The tumulus was most often conceived of as a collective grave of foreign soldiers, casualties of the First World War or one or other of the campaigns of the Balkan Wars. Ironically, that which was there the longest was considered by many as something of an import, a monument of a foreign culture" (PAPADOPOULOS *et al.* 2008: 699).

My final example is the prehistoric tumulus settlement of Assiros, Macedonia, and I shall quote from a remarkable anthropological study of the modern community of Assiros, *Fields of Wheat, Hills of Blood*, by ANASTASIA KARAKASIDOU (1997).

> "This account opens with a prominent point of reference in the local landscape, the prehistoric toumba mount that rises just south-west of the present day settlement of Assiros [...] While British archaeologists have excavated this Bronze and Iron Age toumba extensively, most elderly Assirotes still believe that the mount was [...] constructed by Alexander the Great, the epic ancestor. This they unanimously claimed to have learned from their former schoolmaster. Alexander [...] has become an impassioned present-day national symbol of the Greekness of Macedonia, both past and present" (KARAKASIDOU 1997: 33).

> "Those present-day inhabitants of Assiros who are descendants of Guvezna's [Slav former name of the village] Slavic-speakers conceive of and conduct themselves as nothing less than Greek citizens and full-fledged members of the nation of Hellenes; they maintain strong convictions that Macedonia is Greek and some

believe that they themselves are the Greek descendants of Alexander the Great. Yet they do not call themselves Macedonians. Some may have considered themselves Macedonians in the past and several referred to their former Slavic language as Macedonian (Makedhonika)" (KARAKASIDOU 1997: 22).

Fig. 2a.

"Origin myths regarding Assiros shared a certain timelessness [...] dates are often confused or linear chronology becomes irrelevant as local interpretations [...] hinge on symbols and events rather than on points in time. [...] figures like the Great Alexander belong to both the near and distant pasts" (KARAKASIDOU 1997: 34).

Fig. 2b.

Fig. 3.

Discussion

The rural societies I have cited in this paper appear to deploy ancient objects in the landscape and, occasionally, where there are individuals with higher education and access to specialist libraries within the community, also fragments of academic knowledge they have become aware of, to structure their current existence into a greater sense of importance than their own communal memories can provide. This reflects a general absence of local traditions passing on reliable information regarding the deeper past of the community, something which appears widespread among traditional rural societies, as we see by comparison with a discussion of memory among the non-elite on Malta by an anthropologist:

> "The memory of the populace appears to have little depth, no collective focus, not much dynamism" (SANT CASSIA 1999: 252).

It also recalls the comments of social historian LE ROY LADURIE on the Medieval inhabitants of Montaillou in South-West France:

> "the people of Montaillou lived in a kind of 'island in time', even more cut off from the past than from the future. 'There is no other age than ours', said one. This absence of a historical dimension went with a general use in speech of the present indicative tense without logical connections with past and future" (LADURIE 1978 : 282).

This agrees with our earlier observations on the ways in which memories are transformed or even created to accommodate the 'way we live now', a process which frequently elides the actual passage of time and real alterations in the character of local communities over the medium- to longer-term.

Conclusions

It appears that the intervention of historians and archaeologists into traditional rural communities, encouraged by their most highly-educated members, but also the villagers' own search for anchoring origin myths to take pride in, are attempts to compensate for a general absence of reliable knowledge concerning the true historical trajectory connecting local societies to their long-term past. Tales commonly repeated regarding village history or the significance of local monuments, frequently introduce famous national or even international events or personalities, in order to claim that the community has appeared on the stage of history as somewhere of significance. This is surely a direct reaction to official neglect by nationalist historians of local history at the village level, depriving rural societies of a past they could be proud of in the present. Real memories are often merely three-generations deep. As we found in the summer of 2012, and as Karakasidou found for Assiros village, providing such societies with detailed long-term archaeological and historical information about the origin of their village and its predecessors proves a revelation to many villagers but remains contested by just as many other members of local society.

Working with local villagers usually benefits immensely from the already-existing interest of members with a natural inquisitiveness about their community-history. They can be schoolteachers, lawyers, or retired professionals who have returned from a more cosmopolitan world to their villages of origin. Such individuals increasingly edit community newspapers and even write books about local life and traditions, and are likely to have collected academic articles concerning their district even before direct contact with archaeologists and historians who are researching in the region. Moreover, such individuals with lengthy or even life-long knowledge of their area accumulate knowledge external academics are unaware of, thus creating an ideal mutual research endeavour between these two groups.

On the other hand, the need felt by villagers to possess their own narratives independent of outside interference, which purports to 'tell them who they are and were', is more than understandable when in many Mediterranean countries the state is viewed with suspicion, while as noted, available national histories have generally neglected initiatives to create local histories in the provinces. There is clearly a requirement for external researchers to work sensitively with the communities they embed themselves into, establishing collaborative ties with key individuals within them, who know already much and wish to know more through knowledge-sharing.

On the other hand, I would completely reject the Post-Modern proposition, supported occasionally by theoretical gurus in Archaeology such as IAN HODDER (1999; see BINTLIFF 2011), that all versions of the past are worth equal consideration:

> "it cannot any longer be adequate to separate an objective past defined by archaeologists and a subjective past defined by non-archaeologists. We all interpret the past from diferent perspectives" (HODDER 1999: 200).

Quite so, but as archaeologists or historians we have a professional obligation to attempt as honest a reconstruction as our training and the available evidence permit, to give people in the past their real place in history and pass such insights on to our collaborators within their descendant communities.

The challenge set by the famous German 19th century historian Leopold von Ranke should remain our collective goal, to strive to understand the past "*wie es eigentlich gewesen*" (how it actually was…). The full quote is moving and honest and worth repeating in full:

> *"Man hat der Historie das Amt, die Vergangenheit zu richten, die Mitwelt zum Nutzen zukünftiger Jahre zu lehren, beygemessen: so hoher Ämter unterwindet sich gegenwärtiger Versuch nicht: er will bloss sagen, wie es eigentlich gewesen".* [People have attributed to History the task to judge the Past, to teach it to our contemporaries for their advantageous use in future years: such high aspirations are

far from the effort before you: it merely desires to tell you how it really was then]. (VON RANKE 1824: Preface to Part 1).

References

ALEXANDRI, ALEXANDRA (2002), Names and emblems: Greek archaeology, regional identities and national narratives at the turn of the 20th century. *Antiquity* 76, 191–199.

BARRETT, JOHN C. (1994), *Fragments of antiquity. an archaeology of social life in Britain, 2900-1200 BC.* Oxford.

BINTLIFF, JOHN (2011), The death of archaeological theory? In: John L. Bintliff – Mark Pearce (eds), *The death of archaeological theory?* Oxford. 7–22.

DELTSOU, ELEFTHERIA (2009), Researching biographies of archaeological sites: The case of Sikyon. *Public Archaeology* 8 *[Archaeological Ethnographies]*, 176–190.

FORBES, HAMISH (2009), Researching Ekina ta Khronia [Times Past] in a Rural Greek Community. *Public Archaeology* 8 *[Archaeological Ethnographies]*, 88–107.

FORBES, HAMISH (this volume), It's the fort that counts. Cultural marginalisation and alternative monumentality in a Greek community. In this volume ###-###.

HALL, JONATHAN M. (1997), *Ethnic identity in Greek antiquity*. Cambridge.

HODDER, IAN (1999), *The archaeological process. An introduction*. London.

KARAKASIDOU, ANASTASIA (1997), *Fields of wheat, hills of blood. Passages to nationhood in Greek Macedonia, 1870-1990*. Chicago.

LADURIE, EMMANUEL LE ROY (1978). *Le territoire de l'historien Vol. 2.* Paris.

PAPADOPOULOS, JOHN K. – LORENC BEJKO – SARAH P. MORRIS (2008), Reconstructing the prehistoric burial tumulus of Lofkend in Albania. *Antiquity* 82, 686–701.

SANT CASSIA, PAUL (1999), Tradition, tourism and memory in Malta. *Journal of the Royal Anthropological Institute* 5, 247–263.

VON RANKE, LEOPOLD (1824), *Geschichte der romanischen und germanischen Völker von 1494 bis 1514*. Leipzig – Berlin.

It's the fort that counts.
Cultural marginalisation and alternative monumentality in a Greek community

HAMISH FORBES

Abstract:
Greeks' use of their ancient past for political purposes and as cultural capital, widely recognised in archaeology and anthropology over the last two decades, resides to a great extent in Greece's Classical and Hellenistic archaeological heritage, especially the architectural monuments of Athens. Direct cultural and historical links back to this ancient heritage have been used by those at the centres of power to define 'proper' Greek identity. This situation effectively excludes several ethnic minorities who have inhabited Greece for centuries before its independence but whose existence the Greek state denies. This contribution briefly discusses the problems and penalties that these unacknowledged ethnic minorities continue to face. It then explores how members of one such ethnic minority community use an alternative historical narrative and an alternative monumentality[1] *to connect themselves to a past which allows them to demonstrate their own historical roots as proper Greek citizens and to contest the dominant national(istic) rhetoric on Greek identity.*

Introduction

This article focuses on a small community in Greece which belongs to an ethnic minority whose very existence the Greek state has denied for over a century. Simultaneously, however, this and other similar formally unacknowledged minorities have been the target of regular official and unofficial opprobrium since the 19th century as not being 'proper' Greeks. In the face of such denigration of what they themselves consider to be their unimpeachable Greekness, members of this local community have appropriated material remains (the ruins of a small early 19th century fort) to express their claim to be considered fully and properly Greek despite the rhetoric of others. While these ruins might be considered of minimal archaeological significance in a country which can boast the heritage

1 'Monumentality', a term widely used in archaeology but rarely defined, relates to items of material culture, normally though not exclusively immobile and of substantial size, which have important cultural resonance, normally associated in some way with memory.

of the Acropolis, Delphi, Olympia, Mycenae and Knossos, the local community have monumentalised them and transformed them into a very visible political statement. These ruins connect them to the struggle for the liberation of Greece from the bondage of the Ottoman Empire – an episode celebrated annually with a major national holiday and regularly celebrated over the years by various issues of postage stamps (on the political importance of postage stamps in Greece, see GOUNARIS 2003). In addition, again ignoring the material cultural (ancient Greek) traditions considered essential in the dominant national discourse (GEFOU-MADIANOU 1999: 413), they have constructed another monument to strengthen their links to this same defining period in modern Greece's history.

First, however, a brief digression into terminology and language. In this article I shall use the term 'ethnic minority'. Max Weber defined 'ethnic minority groups' as

> "those human groups that entertain a subjective belief in their common descent because of similarities of physical type or of customs or both, or because of memories of colonization and migration; this belief must be important for the propagation of group formation" (WEBER 1978: 389).

Such a group is one whose origins are other than that (or those) of the dominant majority, but there is no automatic implication in the English phrase that members of such a group cannot have the same nationality or national identity as that of the majority, despite non-standard origins. However, in Greek the word *ethnos*, has significantly different connotations, indicating in a formal sense, 'nation' and nationality. In my experience, a number of Greek people who meet the English term 'ethnic minority' used in discussing Greece have assumed that it indicates a group of people who consider themselves to have a nationality or national identity *other than* Greek: because of this they have considered the topic to be highly contentious. In the discussion which follows, this is the very opposite of the case: the community under discussion wish to emphasise their *Greek* national identity, although via a different route from that presented as 'correct' by those at the nation's centres of power who wish to dominate the national discourse on the 'proper' national identity. GEFOU-MADIANOU (1999) repeatedly uses the term 'marginal' and 'marginalised' for the group which she describes, deliberately not defining them as an ethnic minority (GEFOU-MADIANOU 1999: 413). While they are marginalised, these terms are not quite appropriate in the present context.

Despite the potential for mixed messages, therefore, I shall use the phrase 'ethnic minority' in this article since I cannot think of a better English alternative, but with the understanding that it does not signify any alternative *national* identity.

In presenting this study of contested identities and the use of alternative pasts and their associated monumentalities as self-validation, I have found Gefou-Madianou's work (GEFOU-MADIANOU 1999) to be particularly helpful. In analysing identity-formation in a community of one of Greece's ethnic minority groups, she points out that identity

> "should be understood not as an object that can be defined outright, but rather as an ongoing process whereby relations of power, authority, and authenticity are negotiated and formulated within particular social and political contexts [...]" (GEFOU-MADIANOU 1999: 414).

This approach is very much at the heart of this article. I shall discuss how the use of a particular monument as validation of the community's social worth is part of a group's dialogical process of negotiating its identity with what it views as a culturally and politically dominant Other, which in turn views members of the community as culturally, ethnically and morally inferior Others.

There is no doubt but that Greeks as a nation are immensely proud of their ancient cultural heritage. Over the past two decades a number of archaeologists, social anthropologists and historians of 20th century Greece have drawn attention to the way in which Greeks have used their ancient written and archaeological heritages as cultural capital. Almost from the beginnings of the modern Greek state in the earlier 19th century, modern Greeks' origins in Classical antiquity have formed a key element in constructing their national identity, especially in times of crisis (e.g. CARABOTT 2003; GOUNARIS 2003; HAMILAKIS 2003; HAMILAKIS 2007; HAMILAKIS – YALOURI 1996; MCNEAL 1991; STEWART 2003; SUTTON 1998: 173–8). Governmental and other Greek elites using 'Antiquity' in their rhetoric have regularly referred to the Classical past when considering their identity, as also have political out-groups, such as communists and leftists, who have striven to establish themselves as an alternative elite (see e.g. CARABOTT 2003; HAMILAKIS 2007: 169–204; HAMILAKIS – YALOURI 1996: 119).

A striking recent example of the use of ancient Greece as political capital was a statement by the leader of LAOS, a right-wing nationalist political party, concerning a government initiative to grant citizenship to Greek-born children of immigrants.

As reported by the newspaper *Kathimerini* on 25 January 2010, he stated: "Greece is saying 'no' to this bill because it does not want Hellenism to be diluted. Greece belongs to its history: we were building the Parthenon when they were still living in trees". Variations on this final phrase seem to have been circulating in Greece for some time. STEWART (2003: 485) recounts a variant: "when we were developing mathematics, the English were still hanging off trees".

That the Classical past, particularly its material past as represented by the archaeological record, is at the core of modern Greeks' national identity has become an accepted fact for many archaeologists. Hamilakis very persuasively sets out what has become essentially the dominant discourse on the subject in his recent award-winning book (HAMILAKIS 2007). However, less often discussed is the fact that this idea of a direct link to an ancient Greek past is very much an invented tradition (ANDERSON 1991: 42; GEFOU-MADIANOU 1999: 418; HOBSBAWM 1983: 264). It originated in the 18th century in the rhetoric of educated members of a high status Greek-speaking commercial and seafaring class which had emerged, largely outside of Greece. These intellectuals had 'rediscovered' the Greeks' Classical heritage: as part of their political programme they presented their homeland to the West as the land of descendants of the ancient Greeks, desperate for liberation but oppressed by the infidel Ottoman Turks. Their theme was that the ancient Hellenic culture had continued unbroken through the generations until their own time. In doing so they played a crucial role in the pre- and post-independence intellectual revival of Greece (GEFOU-MADIANOU 1999: 417–418; HAMILAKIS 2007: 75–77). As part of their rhetoric this group claimed principle ownership of the Classical past and, simultaneously, a position of symbolic superiority over other Europeans (HAMILAKIS 2007: 75–77). The emphasis on the ownership of a Classical past as a device for defining those who do not own it as inherently inferior is very much at the heart of this contribution.

In parallel with archaeologists' recognition of the importance of the Classical past for Greek identity, another substantial but entirely separate discourse has been developed during the same time frame – the last two decades – by legal specialists, linguists and anthropologists, both Greek and non-Greek. This relates to communities of Greeks who have lived for centuries in the area that is now Greece from well before there was a Greek state, but whose very existence the Greek state officially denies. These are Greeks who speak languages other than Greek, or whose ancestors spoke them, since many of these languages are dying under the onslaught of government-sanctioned hellenisation programmes. They include Turkish Moslems, Bulgarian Moslems, Aroumanians (Vlachs) speaking a form of Rumanian, and Arvanites speaking a form of Albanian, among others (e.g. BINTLIFF 2003; BRUNNBAUER 1999; DEMETRIOU 2004; GEFOU-MADIANOU 1999; HART 1999; HERZFELD 2003; KARAKASIDOU 1993; KRETSI 2002; LAWRENCE 2010; LIVANIOS 2006; POLLIS 1992; SASSE 1998; STAVROS 1995; and several in TSITSELIKIS 2008: 29 n. 2).

In effect, even in the 21st century the Greek state continues the former Ottoman Turkish practice: 'nationality' based on modern concepts of ethnicity, associated with separate languages and cultures, was not recognised by the Ottoman state (VUCINICH 1962: 605; ABU JABER 1967: 214; GOFFMAN 2002: 9). Instead the Greek state, like the preceding Ottomans, recognises two groups besides Christians – Moslems and Jews – but the existence of any minority groups based on (non-Greek) ethnicity has been denied (HERZFELD 2002: 906). Thus, a United Nations body recently highlighted Greece's poor record on diversity and minority cultures, noting that as of 2011, Greece had still not ratified the Framework Convention for the Protection of National Minorities or signed the European Charter for Minority Languages. Additionally it noted that judgements by the European Court of Human Rights against Greece in 1998, 2007 and 2008, all for violations of Article 11, (concerning freedom of assembly and association, relating to the ability of minority groups to form their own associations) had still not been implemented (UNHCR 2011).

Despite sometimes heavy-handed action by the Greek state over many decades, and considerable levels of discrimination by Greeks who consider themselves 'superior' (e.g. WHITMAN 1990: 17–21) however, non-Greek language-groups still exist in Greece. Nevertheless, those who claim too publically the existence of any *ethnic* alterities within Greece's borders have been punished (e.g. MAGLIVERAS 2009: 23–24). A number of high profile legal prosecutions of individuals who insisted on identifying as belonging to an ethnic minority have occurred in the last 20 years or so.

Thus, in 2001 a Greek citizen belonging to the Aroumanian minority was sentenced to 15 months in jail simply for distributing a leaflet produced by a body linked with the EU listing all the lesser-used languages of Europe, including those spoken in Greece. His crime was "disseminating false information" which could "provoke public anxiety and give the impression that there are minority problems in Greece" (BALTSIOTIS *et al.* 2001: 145–148). Similarly, in 1990 a former parliamentarian and community leader was sentenced to 18 months' imprisonment for disrupting public peace, because he had described the Turkish-speaking Greek citizens of the region as Turkish rather than Greek. The Court of Appeals and the Court of Cassation (*Areios Pagos*) subsequently upheld the sentence (EUROPEAN COURT OF HUMAN RIGHTS 1996: especially section 11; HUMAN RIGHTS WATCH 1999: especially 11–14). A US State Department document on human rights in Greece produced some 20 years after this event evinced concerns about the human rights of this minority, stating that Turkish-speaking Greeks were still legally barred from self-identifying as Turkish, despite repeated European Court of Human Rights decisions supporting their right to do so (WIKILEAKS 2009: especially paragraphs 1–3, 6).

It will be apparent from the above that a focus on a Classical and Hellenistic past as being *the key attribute* to Greek identity has important ramifications for minority groups in a country which was, at the dawn of its independence, part of the Middle Eastern ethnic mosaic after centuries of Venetian and Ottoman rule. It was heterogeneous, polyglot, multi-ethnic and with three separate major religions – Moslem, Jewish and Christian (mostly Orthodox but with a significant Catholic minority [LIVANIOS 2006: 43]) – and its peoples spoke many languages as well as highly varied Greek dialects (SASSE 1998: 41; LIVANIOS 2006; TSITSELIKIS 2008: 28). Linking Greek identity so firmly to a Classical – and especially an *Athenian* – past, particularly in association with an ideology which considers those without such links as inherently inferior, effectively marginalises those citizens whose ethnic origins do not link them directly to it.

This article attempts to view the issue of the marginalisation of these ethnic minority communities from a perspective other than that simply of the impact of the dominant national(istic) discourse of a Classical-Hellenistic past on groups which are best seen as indigenous minority groups, as defined by the ILO *Indigenous and Tribal Peoples Convention* (ILO 1989; see also WATKINS 2005: 430). Instead, the approach taken here follows that of GEFOU-MADIANOU (1999: 412–414): the identities of these indigenous minorities are formed via a dialogical interplay between the members of these groups and the rhetoric of the hegemonic groups who generate the dominant discourse on national identity.

Arvanites

It is against this background that I wish to turn to one of these marginalised groups whose origins are seen as being from outside Greece: Arvanites, who speak Arvanitika – a form of Albanian. As they belong to the majority Greek Orthodox faith they have mostly merged well with the mainstream Greek-speaking population, although largely monoglot Arvanitika speakers existed in some communities in the later 20th century. There are suggestions that they may have settled in some parts of Greece as early as the 9th century, and their presence in Greece in the 13th and 14th centuries AD is well documented (MAGLIVERAS 2009: 15).

Arvanitic communities were widespread on the Greek mainland and on a number of the islands in 1821, at the start of the Greek Revolution (SASSE 1998: 44–46), to which they made a significant contribution. In the Peloponnese there were large numbers of Arvanitic communities, both in the mountains and in the coastal islands of Hydra, Spetses and Poros with their very substantial fleets. It has been argued that, with much of the fighting during the Greek Revolution focused in the region, Arvanites provided the main military muscle which realised the nationalist dreams of the Ottoman Greek merchant class, the prime movers of the independence struggle (LAWRENCE 2010: 5).

Arvanites have not fared well within the Greek state, however. In the mid and later 19th century the Greek state initiated a hellenisation policy towards all unacknowledged linguistic alterity groups. This was intensified towards the Arvanites in the later 19th and early 20th centuries, as a reaction to the rise of Albanian nationalism on Greece's borders (GEFOU-MADIANOU 1999: 420; LIVANIOS 2006: 62–63; SASSE 1998: 50–52). Further bouts of especially intensive oppression, with active discouragement and repression of the use of Arvanitika occurred under the dictatorships of Metaxas (1936–40) and of the colonels (1967–74) (SASSE 1998: 55; GEFOU-MADIANOU 1999: 420–421). Official intolerance towards Arvanitika speakers has resulted in a drastic reduction in their numbers since the later 19th century, from approximately one fifth of Greece's population then to somewhere around 1% of the population now, with regular use confined to a much smaller number (SALMINEN 1999; see also UNHCR 2011). In southern Italy where there *is* a degree of tolerance, there has been a less marked decline in the number of speakers of the equivalent Albanian dialect – *Arbëreshë* (HELSINKI HRW N.D.).

Arvanites' long-term difficulties have been further complicated over the last two decades or so by a flood of an estimated 400–500,000 people from Albania (some 4% of the country's total population [BALDWIN-EDWARDS 2002: 212, 214]), many, if not most residing and working illegally. By the beginning of the new millennium there was a near-hysteria concerning these recent migrants.

> "Greek feelings towards the Albanian immigrants are overwhelmingly negative. Of all the migrants in Greece, [...] the Albanians are the

> least appreciated. The sad truth is that racism and fear against the 'Balkan other', personified by the economically devastated and desperate Albanian immigrant, is an everyday reality for the Greek population. [...] These Greek insecurities cover a wide spectrum, from fears that Albanians might break into houses, stealing and killing, to threats against the cultural and blood 'purity' of the Greek population" (VIDALI 1999).

Under these circumstances Arvanites are unwilling to valorise publicly their own historical roots beyond Greece's borders.

My own experiences as an ethnographer working in an Arvanitika-speaking region of Greece reflect the situation presented in the scholarly literature. I have witnessed ordinary 'mainstream' Greeks haranguing groups of Arvanites, stating that they are essentially Turks (the traditional enemy of the Greeks to this day) – a view also reported by MAGLIVERAS (2009: 31) who records the phrase "Turkish mercenaries" applied to Arvanites. I have also been harangued publicly at an academic conference by a senior Greek academic who complained that I and a number of colleagues who happened to be studying Arvanitic communities had no right to do so because Arvanites were not 'proper' Greeks.

The Arvanitic identity is complex. In my experience, despite the intolerance towards them demonstrated by those in the centres of cultural and political power, Arvanites were very proud of their Arvanitic heritage and more than happy to present it to those outsiders whom they considered sympathetic. At the same time though, they are also proud to identify themselves as being unimpeachably proper Greek citizens. They recognise that the Greek Arvanitic identity is quite separate from that of Albanian nationals residing in Greece, yet I have also witnessed the fact that some – especially those whose Greek was relatively limited – were happy to converse with Albanian migrants in their own language (see also BINTLIFF 2003: 139; GEFOU-MADIANOU 1999: 416). The ethnographic literature describes some Arvanitic communities that have developed special relationships with immigrant Albanians via a recognition of shared historical origins (ATHANASSOPOULOU 2005; MAGLIVERAS 2009). However, because of the kinds of prejudice described here and in the ethnographic literature (e.g. GEFOU-MADIANOU 1999: especially 415–416; MAGLIVERAS 2009: 23–24) Arvanites very rarely voluntarily publicised their alternative identity to strangers whose sympathies they could not guarantee, although LAWRENCE (2010) presents a case study in which certain members of one community aggressively flaunted it. Nevertheless, once it was established that I was sympathetic to their minority identity they celebrated it to me, while simultaneously insisting on their *Greek* nationality.

Among themselves Arvanites have traditionally spoken Arvanitika: this is an obvious identity-marker. Even those who speak little of the language, however, often know a number of traditional songs in the language. These songs are important for self-identification *as* Arvanites – as indicated by the numbers of performances available via the internet. Again, however, these tend not to be performed in public arenas, but in more private contexts in which they can be employed for self-definition – see e.g. GEFOU-MADIANOU (1999: 416) for an excellent example from the countryside around Athens. In my experience, Arvanitic songs were definitely *not* sung in public celebrations, such as when bands played and sang at annual saints' day festivals (*paniyiria*), even in communities where Arvanitika was regularly spoken in most homes. They were regularly sung, however, during wedding festivities – and in the secular rituals leading up to them – when everyone could be sure that they were among purely relatives and family members.

The case study

The previous section provides the background to a case study examining how one Arvanitis minority community has appropriated an alternative (non-Classical) monumentality to self-identify as worthy citizens in the face of generations of marginalisation, while simultaneously maintaining their particular identity and engaging in apparent small symbolic acts of defiance against the hegemonic centre. The non-Classical monument in question, of no great intrinsic archaeological significance, has acquired special meaning inasmuch as it facilitates the dialogical interplay between the community's members and the elite groups who generate the dominant discourse on national identity, using an alternative historical tradition also valorised by those in the cultural and political centres of power.

The inhabitants of the small peninsula of Methana, on the southwestern edge of the Saronic Gulf (*Fig. 1*), are part of what was traditionally an Arvanitika-speaking area which includes large parts of the Argolic peninsula, to which it is attached by a narrow isthmus. The peninsula, less than 10 km in diameter, contains ten small villages and a spa resort founded in the late 19th century, which is the administrative centre. As recently as the 1970s most inhabitants engaged in largely self-sufficient farming, supplemented by sales of small amounts of agricultural produce and, for some, a certain amount of wage labour, especially

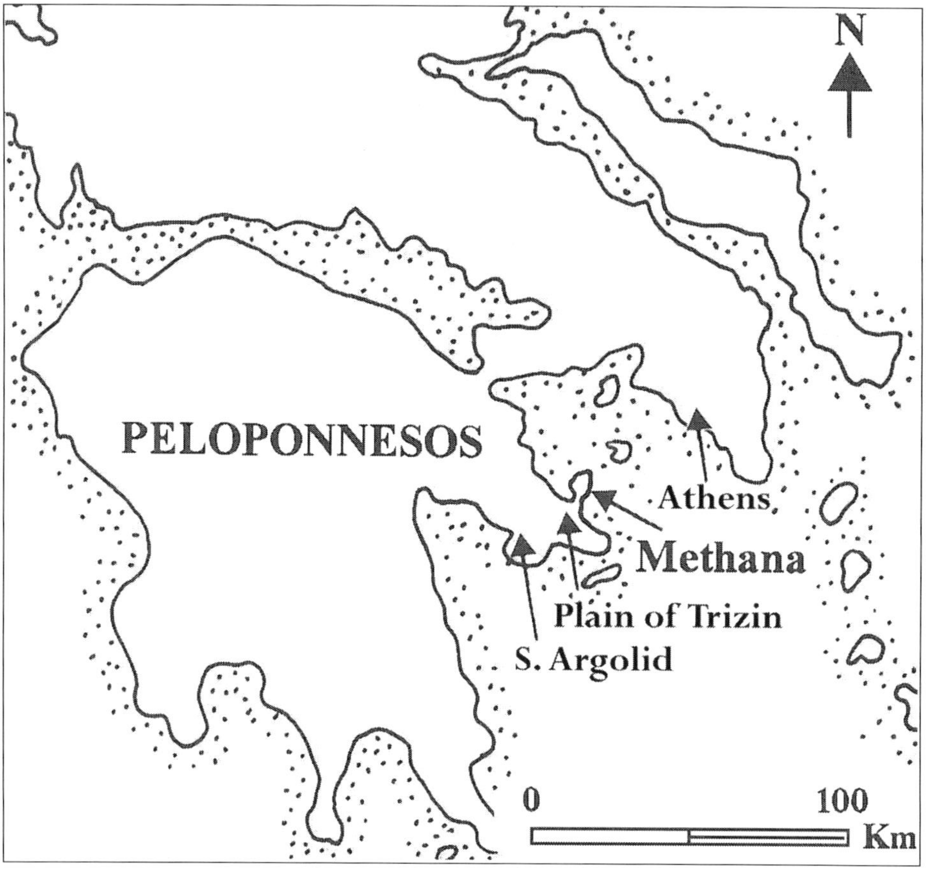

Fig. 1. Methana: location map.

in the spa resort during the busy summer months. The peninsula itself is mostly volcanic in origin, which explains its exceptional ruggedness and the fertility of its soil: its steep terraced hillsides rise from the sea to a maximum height of just under 750m (FORBES 1982).

At the time of my original ethnographic fieldwork, conducted over two years between 1972 and 1974 (FORBES 1982), Arvanitika was the primary language spoken in many homes, especially among older residents. A number of women neither spoke nor understood Greek, and many older people spoke grammatically incorrect Greek. I conducted further research on the peninsula for several summers in the earlier 1980s, co-directing an archaeological survey (MEE – FORBES 1997) during which time I continued with a certain amount of ethnographic research.

During the same period I also conducted ethnographic research in the Southern Argolid, another Arvanitika-speaking area nearby on the Peloponnesian mainland (FORBES 2000). I spent a further period of ethnographic research on Methana during the summer of 1998.

On my way to Methana in 1998 I took a taxi in Athens. The driver harangued me over a recent scandal concerning the Elgin Marbles. I then realised that through all the years I have been working on Methana and in the southern Argolid, local people have never commented on the ancient monuments in Athens, still less on the Elgin/Parthenon Marbles controversy. While it might be thought that this lack of interest resulted from a rural population disdaining city life, this does not seem to be the answer.

There are some notable ancient sites on the peninsula of Methana, especially the acropolis of the ancient city of Methana (MEE *et al.* 1997: 122–127). It might be expected that Methanites would feel some kind of historical connection between themselves and *their* local monuments, especially in light of the dominant archaeological discourse on the importance of the materiality of the ancient past for Greek identity. However, Methanites do not associate themselves historically with an ancient Classical and Hellenistic past – a situation reported also in another Arvanitic village elsewhere in the eastern Peloponnese (DELTSOU 2009). For Methanites, ancient and prehistoric sites are vaguely interesting, but in many ways no more – and possibly even less so – than natural wonders on their peninsula such as the most recently erupted volcanic cone or a cave discovered only a few decades ago (FORBES 2009: 101). Despite not connecting themselves to a glorious Classical-Hellenistic past, Methanites considered themselves to be every bit as Greek as all other Greek citizens, and much better Greeks than some others of the nation, who in Methanites' opinion have nasty habits, like knife-carrying and indulgence in blood feuds – things to be deplored among proper citizens.

There is one archaeological site which Methanites do not ignore: it has a very special place in how they identify themselves *as* Methanites and simultaneously connects them to unimpeachable Greekness. In 1826, during the Greek War of Independence, the French philhellene Charles-Nicolas Fabvier needed a safe place in which to train his inexperienced troops. He chose the Methana isthmus because of the safety offered by the ruggedness of the peninsula and the narrowness of the isthmus joining it to the mainland (ST. CLAIR 1972: 191–192). Little remains of the settlement and military base that he built on the isthmus and named Takticopolis, apart for the ruins of a small fort (MEE *et al.* 1997: 165).

Although the site is now called the Kastro Favierou (Fabvier's castle) the original name lives on, now being applied to a neighbouring village. In the post World War II period the Greek government forced the villagers to change its original Albanian name to a Greek one as part of its programme of cleansing ethnically non-Greek place-names from the map. The villagers therefore chose the name of Fabvier's settlement, again emphasising the importance of the site to the local community.

Over several decades Methanites have emphasised to me the importance of this site as a monument and an indicator of their community's contribution to the founding of the nation. This pride is evident in the ruins themselves. A large Greek flag was painted on it many years ago (*Fig. 2*). The date is uncertain but this design was superseded in 1978: a second version painted nearby (*Fig. 3*) uses the current design (BRESCHI N.D.). Another much smaller painting of the Greek flag – yet another design – has been painted inside a gun-slit (*Fig. 4*). In the 1980s, when I mentioned to Methanites the initial surmise of the Methana Archaeological Survey team that the main fortification was Venetian they were

Fig. 2. The Kastro Favierou*: Charles Fabvier's Revolutionary War fortification on the Methana isthmus.*

Fig. 3. Two Greek national flags prominently displayed on the Kastro Favierou.

very disappointed. While a Venetian date might be considered more interesting by archaeologists, for Methanites it was specifically as a War of Independence monument that the ruins had their full meaning.

As previously noted, the military dictatorship of 1967–1974 saw increased repression of Arvanites. During this period Methanites felt it necessary to construct a second War of Independence monument to bolster their links with that crucial period of the nation's formation. This was a memorial stele for the Methanites who fought in the War of Independence, especially the leader of the Methana contingent, who fell in battle on behalf of the cause (*Fig. 5*). While the idea of a memorial stele has its roots in ancient Greece, the motifs employed in this one eschew any such connection: in fact the inclusion of a Christian cross in the design negates any link to a pagan Classical-Hellenistic past. The monument has been set up close to a church which represents pan-Methana feeling most intensely: it is the focal point where Methanites from all over the peninsula and beyond gather together in large numbers on one day every year for Greece's

Fig. 4. Greek flag-design used during the Revolutionary War, in a gun-slit of the Kastro Favierou.

celebration of the nation's patron saint, St. George (FORBES 2007: 263–264, 370–373).

The importance of these two monuments for Methanites' assertion of being worthy citizens was reified while I was writing up this article. On 26 March 2013, a Methanitis journalist uploaded a blog in an online organ describing itself as the first portal for Piraeus (the administrative centre for Methana's region) and shipping matters (ATHANASIOU 2013). Having described the first ever flag-raising at the fort he tells the fort's history, emphasising that this was where Greece's regular army was born, and the central part it played during the War of Independence. He then proceeds to mention the contingent of Methanites and their involvement in a series of iconic battles under famous Revolutionary War leaders, going out of his way to note that the leader's name was an Arvanitic one, while again emphasising his ultimate sacrifice in battle. These monuments, therefore, one original and one retro-constructed, rather than the ancient city ruins, allow Methanites to identity as worthy Greek citizens by associating their place with a glorious past via the

Fig. 5. Stele commemorating the leader of a Revolutionary War band of Methanites, set up in 1968.

historic events which founded the nation. They likewise demonstrate Methanites' status as true patriots despite constant denigration by those in ethnic, class and political power-centres.

Finally, Methanites identify themselves as thoroughly Greek yet simultaneously in contestation with those in the centres of power. It is hard to disentangle the extent to which this bipolarity reflects their identity as unacknowledged Others and how much it demonstrates that as the anthropologist Russell Bernard has stated, "to rural Greeks, one overwhelming social fact is the 'specialness' of the *topos*, the *place* where they live" (BERNARD 1976: 289). The painted flag semi-hidden in the embrasure of a gun-slit (*Fig. 4*) may be relevant here. Superficially it is something of a mystery. Unlike the two large and prominent flags mentioned previously it is small, and semi-hidden in a gun-slit on the interior of the fort:

not a ringing endorsement of patriotic pride. Yet it may have a symbolism and significance disproportionate to its size and position.

During the Greek War of Independence several different flag designs were in circulation concurrently. The design painted here was especially popular in the Peloponnese. It had a particular association with one of the greatest war leaders of the revolution, himself a Peloponnesian, Theodhoros Kolokotronis (BRESCHI N.D.; GREECE FLAGS N.D.). Since the war band mentioned on the stele served under his nephew, Nikitaras, a connection to Kolokotronis seems logical.

While it might seem unremarkable that a Peloponnesian Revolutionary War site should be associated with a flag associated with that place and time, I believe there are messages of contestation in its use. Significantly, the flag displayed is *not* the design adopted officially in the first Greek National Assembly in 1822. In other words, the design used here is contrary to the official one of that time. Its small size and semi-secretive positioning are also suggestive of an unofficial act of rejection of the dominant discourse. Kolokotronis, who used this design, was first and foremost a soldier, not a politician. He was both one of the most influential individuals of the Revolutionary War and also renowned for being regularly at loggerheads with the politicians in the new Greek government. Furthermore there is a widespread belief in Greece that he had Arvanitis origins, as indicated by a burgeoning Greek metaliterature on the internet.

In addition, against all geographical logic, Methana and a narrow adjoining strip of the Peloponnese are administered from the capital, Athens and Piraeus, not the rest of the Peloponnese. Some Methanites have expressed their antipathy for this arrangement since they are isolated administratively from their immediate neighbours, and local government policy in this rural location is dominated by the needs of the capital city.

In other words, superficially the painting of this little graffito can be read as a straightforward patriotic statement linked to the time-frame of the fort. At another level, however, it can be read as a minor act of defiance, in a symbolic language understandable to all Greeks, against cultural hegemonisation by those at the centres of power: a statement of Methanites' separate regional identity and also a re-affirmation of their contribution as Arvanites, to Greek national society and history.

Conclusion

This article is about identity-formation and how different pasts and traditions are used in the process. Methanites, as part of an Arvanitic minority use language and songs as part of the way in which they perform their identity among themselves *as* Arvanites. However, in the dominant ideology developed by the Greek nation's intellectual elite, anyone whose cultural roots do not link directly to a Classical past is considered inferior. This definition of superior/inferior effectively marginalises all those groups which have resided in Greek lands for centuries but whose roots self-evidently do *not* link with this preferred past. I have illustrated how this particular minority attempts to resist and contradict their categorisation by the elite as inferior citizens. As part of this dialogical process with a dominant Other this community looks not to the Classical past, but to a monument relating to an alternative period of the past which is also valorised by those at the cultural and political centres of power. This dialogical process with an Other in the nation provides opportunities for performing acts which allow them to represent themselves as 'proper' citizens with a 'proper' and meaningful past.

Acknowledgements

I would like to thank Ömür Harmanşah and the Joukowski Institute of Brown University (Providence RI, USA) for their invitation to give a lecture in April 2009, and Yannis Hamilakis and Aris Anagnostopoulos for their invitation to participate in an Archaeological Ethnographies Workshop on the island of Poros, Greece, in May 2009: a number of the ideas presented here derive from those occasions. Much of the subsequent research on the topic was conducted while I was a Visiting Professor at Aarhus University, Denmark: warm thanks are therefore due to that university and especially the Institute for History and Area Studies for the invitation to such a pleasant and fruitful research environment. I also wish to thank Elisabeth Niklasson and Thomas Meier for the invitation to participate in their session at the EAA meetings in Helsinki and their subsequent invitation to submit this contribution for their final publication. I owe Linos Papachristou a particular debt of gratitude for his kind permission to publish the illustrations in *figures 3* and *4* and for his efforts to ensure a high enough resolution. None of the above, however, is in any way responsible for any infelicities in this work, nor the views expressed therein.

References

Abu Jaber, Kamel (1967), The millet system in the nineteenth-century Ottoman empire. *The Muslim World* 57, 212–223.

Anderson, Benedict (1991), *Imagined communities.* London – New York.

Athanasiou, Savvas (2013), *Τιμή στους αγωνιστές του κάστρου του Φαβιέρου.* Portnet.gr. (http://www.portnet.gr/component/content/article/42-saronikos/12550-timh-stous-agonistes-tou-kastrou-tou-favierou.html [31.03.2013]).

Athanassopoulou, Angélique (2005), 'Nos albanais à nous': travailleurs émigrés dans une communauté arvanite du Péloponnèse. *Ethnologie Française* 102, 267–278.

Baldwin-Edwards, Martin (2002), Semi-reluctant hosts: southern Europe's ambivalent response to immigration. *Brown Journal of World Affairs* 8.2, 211–229.

Baltsiotis, Lambros – Leonidas Embiricos – Maria Margaronis (2001), Speaking in tongues. *Index on Censorship* 30, 145–151.

Bernard, H. Russel (1976), Kalymnos: the island of the sponge fishermen. In: Muriel Dimen – Ernestine Friedl (eds), Regional variation in modern Greece and Cyprus: Toward a perspective on the ethnography of Greece. *Annals of the New York Academy of Sciences* 268, 289–290.

Bintliff, John (2003), The ethnoarchaeology of a 'passive' ethnicity: the Arvanites of Central Greece. In: Brown – Hamilakis 2003: 129–144.

Breschi, Roberto (n.d.), *Bandiere* (http://www.rbvex.it/grecia.html [accessed 04.04.2013]).

Brown, Keith – Yannis Hamilakis (eds) (2003), *The usable past: Greek metahistories.* Lanham.

Brunnbauer, Ulf (1999), Diverging (hi)stories: the contested identity of the Bulgarian Pomaks. *Ethnologia Balkanica* 3, 35–50.

Carabott, Philip (2003), Monumental visions: the past in Metaxas' *Weltanschauung*. In: Brown – Hamilakis 2003: 23–38.

Deltsou, Eleftheria (2009), Researching biographies of archaeological sites: the case of Sikyon. *Public Archaeology* 8, 176–190.

Demetriou, Olga (2004), Prioritizing 'ethnicities': the uncertainty of Pomak-ness in the urban Greek Rhodoppe. *Ethnic and Racial Studies* 27, 95–119.

European Court of Human Rights (1996), *In the case of Ahmet Sadik* v. *Greece*; number 46/1995/552/638. (http://www.cilevics.eu/minelres/coe/court/AhmetSadik.htm [03.04.2013]).

Forbes, Hamish (1982), *Strategies and soils: technology, production and environment in the peninsula of Methana, Greece*. Ann Arbor.

Forbes, Hamish (2000), Landscape exploitation via pastoralism: examining the "landscape degradation" *versus* sustainable economy debate in the post-medieval Southern Argolid. In: Paul Halstead – Charles Frederick (eds), *Post-glacial environmental change in the Aegean area*. Sheffield. 95–109.

Forbes, Hamish (2007), *Meaning and identity in a Greek landscape: an archaeological ethnography*. Cambridge.

Forbes, Hamish (2009), Researching *ekina ta khronia* [times past] in a rural Greek community. *Public Archaeology* 8, 88–107.

Gefou-Madianou, Dimitra (1999), Cultural polyphony and identity formation: negotiating tradition in Attica. *American Ethnologist* 26, 412–439.

Goffman, Daniel (2002), *The Ottoman Empire and early modern Europe*. Cambridge.

Gounaris, Basil (2003), The politics of currency: stamps, coins, banknotes and the circulation of modern Greek tradition. In: Brown – Hamilakis 2003: 69–84.

Greece flags (n.d.), *Greece Flags: Information about the flags and national emblems of Greece.* (http://www.greeka.com/greece-flags.htm [04.04.2013]).

Hamilakis, Yannis – Eleana Yalouri (1996), Antiquities as symbolic capital in modern Greek society. *Antiquity* 70, 117–129.

Hamilakis, Yannis (2003), "Learn History!" Antiquity, narrative and history in Greek educational textbooks. In: Brown – Hamilakis 2003: 39–68.

Hamilakis, Yannis (2007), *The nation and its ruins: antiquity, archaeology and national imagination in Greece*. Oxford.

Hart, Laurie Kain (1999), Culture, civilization, and demarcation at the northwest borders of Greece. *American Ethnologist* 26, 196–220.

Helsinki HRW (n.d.), *Arvanitika*, Report of Helsinki Human Rights Watch. (http://www.greekhelsinki.gr/english/reports/arvanites.html [03.04.2013]).

Herzfeld, Michael (2002), The absent presence: discourses of crypto-colonialism. *The South Atlantic Quarterly* 101, 899–926.

HERZFELD, MICHAEL (2003), Localism and the logic of nationalistic folklore: Cretan reflections. *Comparative Studies in Society and History* 45, 281–310.

HOBSBAWM, ERIC (1983), Mass producing traditions: Europe, 1870–1914. In: Eric Hobsbawm – Terence Ranger (eds), *The invention of tradition*. Cambridge. 263–307.

HUMAN RIGHTS WATCH (1999), *Greece. The Turks of Western Thrace*, Reports Vol. 11, No. 1 (D) (http://www.hrw.org/reports/pdfs/g/greece/greec991.pdf [03.04.2013]).

ILO (INTERNATIONAL LABOUR ORGANISATION) (1989), *Introduction to ILO convention no. 169* [Indigenous and tribal peoples convention] (http://www.ilo.org/public/english/region/ampro/mdtsanjose/indigenous/intro169.htm [03.04.2013]).

KARAKASIDOU, ANASTASIA (1993), Politicising culture: negating ethnic identity in Greek Macedonia. *Journal of Modern Greek Studies* 11, 1–28.

KRETSI, GEORGIA (2002), The secret past of the Greek Albanian borderlands. Cham muslim Albanians: perspectives on a conflict over historical accountability and current rights. *Ethnologia Balkanica* 6, 171 – 195.

LAWRENCE, CHRISTOPHER (2010), Miserable banditti: the pot-head outlaws of rural Greece. *Dialectical Anthropology* 35.1, 33–47.

LIVANIOS, DIMITRIS (2006), The quest for Hellenism: religion, nationalism and collective identities in Greece (1453–1913). *The Historical Review/La Revue Historique* 3, 33–70.

MAGLIVERAS, SIMEON (2009), *The ontology of difference: nationalism, localism and ethnicity in a Greek Arvanite village*. Durham E-Theses Online. (http://etheses.dur.ac.uk/248/ [03.04.2013]).

MCNEAL, RICHARD A. (1991), Archaeology and the destruction of the later Athenian Acropolis. *Antiquity* 65, 49–63.

MEE, CHRISTOPHER – HAMISH FORBES (eds) (1997), *A rough and rocky place: the landscape and settlement history of the Methana Peninsula, Greece*. Liverpool.

MEE, CHRISTOPHER – HUGH BOWDEN – LIN FOXHALL – DAVID GILL – THEODORE KOUKOULIS – GARY TAYLOR (1997), Catalogue of sites: In: Mee – Forbes 1997: 118–210.

POLLIS, ADAMANTIA (1992), Greek national identity: religious minorities, rights and European norms. *Journal of Modern Greek Studies* 10, 171–195.

SALMINEN, TAPANI (1999), *Unesco red book on endangered languages: Europe.* (http://www.helsinki.fi/~tasalmin/europe_report.html #Arvanitika Albanian [27.03.2013]).

SASSE, HANS-JÜRGEN (1998), Arvanitika: The long Hellenic centuries of an Albanian variety. *International Journal of the Sociology of Language* 134, 39–68.

STAVROS, STEPHANOS (1995), The legal status of minorities in Greece today: the adequacy of their protection in the light of current human rights perceptions. *Journal of Modern Greek Studies* 13, 1–32.

ST. CLAIR, WILLIAM (1972), *That Greece might still be free. The philhellenes in the war of independence*. London.

STEWART, CHARLES (2003), Dreams of treasure: temporality, historicization and the unconscious. *Anthropological Theory* 3, 481–500.

SUTTON, DAVID (1998), *Memories cast in stone: the relevance of the past in everyday life*. Oxford – New York.

TSITSELIKIS, KONSTANTINOS (2008), Minority mobilisation in Greece and litigation in Strasbourg. *International Journal on Minority and Group Rights* 15, 27–48.

UNHCR (2011), World Directory of Minorities and Indigenous Peoples – Greece: Overview. October 2011, Minority Rights Group International (http://www.refworld.org/docid/4954ce6a22.html [28.03.2013]).

VIDALI, MARIA (1999), Living in a policy vacuum: the plight of Albanian immigrants in Greece. *Central Europe Review* 1/21. (http://www.ce-review.org/99/21/vidali21.html [28.03.2013]).

VUCINICH, WAYNE (1962), The nature of Balkan society under Ottoman rule. *Slavic Review* 21, 597–616.

WATKINS, JOE (2005), Through wary eyes: indigenous perspectives on archaeology. *Annual Review of Anthropology* 34, 429–449.

WEBER, MAX (1978) [first published 1922], *Economy and society*. Berkeley, CA.

WHITMAN, LOIS (1990), *Destroying ethnic identity. The Turks of Greece*. New York.

WIKILEAKS (2009), CABLE REFERENCE ID: #09THESSALONIKI24, aka Wikileaks id #204217, 27 April 2009 (http://www.cablegatesearch.net/cable.php?id=09THESSALONIKI24 [04.04.2013]).

Disarmed post-socialist archaeologies? Social attitudes to interpreting the past – an interim report from Hungary[1]

ESZTER BÁNFFY

Abstract:

During the decades of socialism in Hungary, theory was something reserved for highly academic persons, having little to do with presenting excavation materials that could tell locals something about their past. Instead, presentation was restricted to rather monotonous museum exhibitions with mainly pottery and other objects in the showcases. In the late 1980s, due to the first micro-regional research programs and the spread of landscape-scale archaeology, this situation changed slightly, resulting in some exhibitions of a higher public interest. However, it was not until after 1989 that major changes started to happen in Hungarian archaeology. One such change was the intensive but often casual introduction of investigations and theoretical strands originating from the natural sciences. Another one was the 1990–2010 large-scale preventive excavations, which brought with them an increased interest in the interpretations and demonstrations of archaeological findings. However, due to a series of changes in legislation, these large scale projects – along with the entire heritage protection scheme – actually became endangered. Furthermore, a third significant change was the unexpected and aggressive spread of fringe archaeological ideas, often ultra-nationalistic, accusing professional researchers of hiding the "truth". These groups are often much more intensively present in society and possess more effective means of communication (via internet) than

1 The author of this chapter has been working with archaeological research in Hungary for three decades. As a field archaeologist and author of several theoretical works, she became a member of, and later on also the head of the scientific advisory board of KÖSZ – Field Service for the Cultural Heritage. As part of this state-funded organisation, which existed from 2007 to 2010, she and her colleagues often raised their voices against laws that endangered archaeological sites. The KÖSZ scientific advisory board introduced mandatory scientific evaluation of the large scale preventive projects. After 2010, most of these achievements were abolished along with the institutions that were meant to secure the archaeological heritage. Apart from this major loss of influence in national heritage management, prehistorians are also often attacked by outsiders when they refuse to accept the ideas of fringe and ultra nationalist groups. As a consequence, the author is deeply involved in current struggles over interpretations of Hungary's past and can by no means be described as a mere observer, making the text both personal and professional, and by necessity also political.

professionals, and thus their "theories" have come to overwhelm the public knowledge about our past. Following these developments, this paper focuses on the motivations behind distorted or harmful nationalist archaeologies in Hungary, both in connection to the social changes of the last decades and the possibilities to increase public outreach of scientific and critical (rather than explicitly romantic and nationalistic) interpretations of the past.

Introduction

A series of works have analysed the specific features that shaped the archaeologies of the countries within the Soviet bloc in various ways. While stressing the differences in the extent of political pressure and limitations of the freedom of research within these countries – according to the general rate of freedom in each state including the Soviet Union itself – there seems to be one feature they all have common: the restrictions put on the exchange of information (BEHRENS 1984; MILISAUSKAS 1986; 1990; KAISER 1995; KUNA – VENCLOVÁ 1995, BIEHL *et al.* 2002, all with further literature). As a consequence, archaeologists studied, worked and published not so much for the outside world, but rather for themselves. Because of this, the last years of the 1980s and especially the changes after 1989 led to an almost traumatic realisation of how much archaeological thought, strands, theories and information they now had to catch up with. This happened almost in parallel with the launch of several labour-intensive large-scale excavations starting right after 1990. In the meantime, and with the benefit of hindsight, it seems as if "the spirit in the bottle" was let free, paving the way for both new knowledge and false doctrines. In the highly complex twenty years that passed after the fall of socialist systems, a series of profound changes took place, many of which needs to be discussed in connection to the development of narratives in archaeology and cultural heritage. Among these I have chosen to pick up three specific themes, each of which, to my mind, effectively highlights and hopefully alerts members of the international scientific community as to the recent discursive changes in Hungarian archaeology, involving threatening developments, for archaeological sites and professionals as well as for Hungarian society at large. While the three themes are discussed separately, the reader will find that they are all related to each other, not least because they are all phenomena that grew to become significant after the fall of the iron curtain.

One such phenomenon concerns archaeological excavations, which from this time on came to consist almost exclusively of rescue works. The new legislation introduced to deal with the huge amount of novel findings and information, both practically and concerning their role in a re-interpretation of the National cultural heritage protection scheme, became highly problematic. In the end, the new legislation only managed to cause even further harm. The second issue has to do with the peculiar development of research, of how the natural scientific strands developed swiftly, while works within more interpretative approaches remained sparse. And lastly, I shall focus on the challenges posed by wilfully distorted or falsified archaeology in Hungary: on the immense spread of fringe or unscientific archaeological narratives, which are now increasingly accepted as real alternatives to professional research. As I shall try to point out, there is a fundamental difference between these escalating beliefs and the various groups involved in so-called Western European "New Age" archaeology. In a Hungarian context, directly misleading archaeological appropriations have now become a serious ideological and political, but above all social challenge, contributing to a collective crisis of identity. In this context I shall also draw attention to a possible communication failure coming from the professional side.

Development-led, large-scale excavations and their consequences

The enormous amount of large-scale excavations is the first striking change to be mentioned. During the first decades of socialist times, smaller scale archaeological surveys, both planned and rescue excavations, allowed for possibilities to process, inventory and store the finds – even if the desired scientific evaluation often fell behind. From the beginning of the 1990s, motorway projects were initiated across the whole country, as well as investments in new shopping malls, pipelines or logistic centres, leading to the sudden increase in flood control and mining activity. These large-scale projects were all consequences of the fall of the Iron curtain. Especially after 2004, when Hungary joined the EU and thus began to receive EU funding for new investments, this meant millions of square meters for us archaeologists to excavate, process and store, year by year.

Initially, these fieldwork campaigns were under the control of the county museums, whose archaeologists carried out the excavations and were supposed to finalise the process, publish the findings and exhibit them to the public. The fact that the documentation systems were not unified caused difficulties, especially in cases when for example a motorway crossed more than one county. In 2007, in order to improve this situation, the Cultural Heritage Field Service (KÖSZ) was established. KÖSZ was a national institution, responsible for the coordination of all preventive archaeological excavations in the whole country. In spite of some problems, mainly due to its hasty formation and to the museum leaders' reluctance to hand over control (and their financial means) to the new Field Service, it nevertheless started to function. Soon the criteria for excavating and documenting the finds and field information, as well as the anthropological and zooarchaeological (etc.) protocols, became unified in the whole country. Apart from the fact that evaluations came to be published in reports and monographs,[2] it seemed that steps were also taken towards creating better solutions for concurrent problems in legislation. For instance, our suggestion to introduce a small regular tax, in order to cease conflicts between excavators and developers, was discussed in several meetings with governmental officials and was later published (Bánffy – Raczky 2010) (*Fig. 1*). Yet, the proposal had no time to be brought into force, as the demolishment of the Field Service was among the first measures to be carried out by the new government in 2010. Fortunately, the laboratories and some of the employed archaeologists were able to continue their work as a department of the Hungarian National Museum.

Meanwhile, the legal rights to carry out large-scale excavations, and more importantly – the money to go with it, were given back to the regional museums. However, as it turned out, the once again heightened expectations of county museologists were crushed by two further happenings. Partly because of the European economic crisis and partly because of the rapidly decreasing international prestige of the new government, today hardly any EU money arrives for new constructions. Due to several new territorial and financial rules, the traditional system based on regional museums is at risk of collapse. Their new budget now ascribed to towns and cities instead of the former counties, may not allow for anything but the cost of storing the finds. As a result, exhibitions may be closed and museum staff dismissed. In order to keep their employments, some museums have already made plans to replace exhibition guards and technical staff with museologists.[3]

---

[2] Under the umbrella of the KÖSZ Scientific Committee and the editorial board a new series was founded: VIA – Monographia Minor in Cultural heritage (series editor: Eszter Bánffy, editor: Judit Kvassay). Since 2008 three volumes are already published, two more are in press and yet another is currently in preparation.

[3] These types of replacements are planned in three large Budapest museums: the Hungarian National Museum, the Natural History Museum and the Museum of Applied Art (http://www.nepszava.hu/articles/article.php?id=494125 [24.11. 2011]). A similar

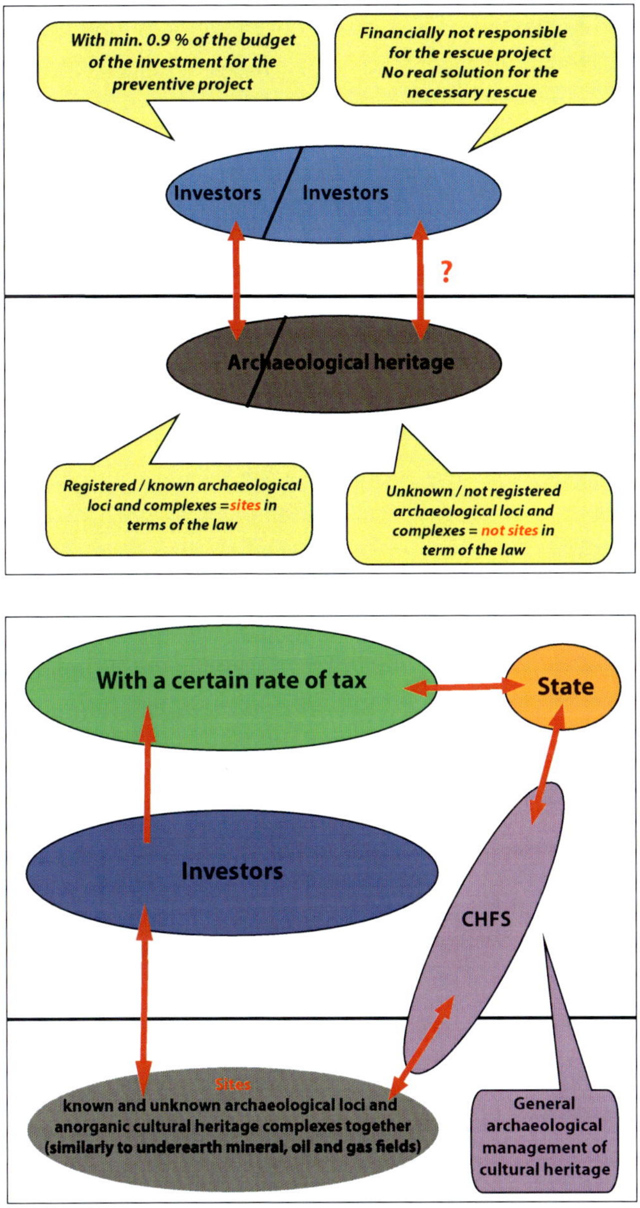

a

b

Fig. 1. Hungarian preventive archaeology and cultural heritage: changes in legislation? a. The state of preventive archaeology according to the proposed legislation in 2009 (after BÁNFFY – RACZKY *2010). b. A proposition for improving the distribution of costs in Hungarian preventive archaelogy (after* BÁNFFY – RACZKY *2010)*

In the meantime, archaeological fieldwork processes have often been accused of being far too slow or unsuccessful, all while the media has continued to portray archaeologists as corrupt and eager for money. This reputation for tardiness, combined with the wretched caricature of the greedy archaeologist, has certainly not helped to create a better communication with the public.

In the past few years, legislation processes concerning cultural heritage in Hungary is a story as long as it is sad.

The first out of the recent modifications made to the Cultural Heritage Protection law,[4] brought with it a dramatic reduction of protection measures relating to large-scale cultural heritage investments. No archaeological organizations or institutions were consulted during the drafting of this legislation. However, after the planned modifications were made public in October 2012, the Association of Hungarian Archaeologists (the only craft union for the archaeological profession in Hungary),[5] prepared a detailed risk assessment analysis with proposed changes and handed it over to the Ministry of National Resources, together with an offer of professional assistance.[6] No response was given by the politicians.

The most critical issue in the accepted legislation is the new limitations set on the excavations. Test and follow-up excavations are now limited to a period of 30 days times two, an extremely restricted period of time which is not meant for each site but for the entire area to be developed, often with many extended sites involved. The second critical issue in this legislation is the radical decrease of money allocated to the necessary excavations taking place in connection to large-scale state or EU funded construction projects. Taking into account the average site density in the intensively settled Carpathian basin, this would mean that not more than 5% of the archaeological sites in the development area could

---

4 Law no. 2001/LXIV, passed through the Hungarian Parliament in 7 November 2011 (law no. 2011/CXLIX).

5 For further information visit the association on facebook or www.regeszet.org.hu. The Association of Hungarian Archaeologists prepared a proposal entitled "A unified concept of development-led archaeological service" that intended to reconcile the differing interests of the investors and cultural heritage protection. Gábor Lassányi, the newly elected president of the association, also reports about their efforts in a blog (http://sirasok.blog.hu [23.04.2012]).

6 The draft, which also contained suggestions for revisions, was prepared by Attila Gyucha, Csilla Zatykó and Eszter Bánffy, May 6, 2012.

be excavated. The remaining 95 % would be demolished, and even the excavated part would end up with hardly any standard documentation or evaluation.[7]

The new legislation does not only undermine the protection of archaeological heritage, but contrary to the declared governmental intention behind the law, it goes against the interests of investors too, and may therefore lead to legally debatable situations (BÁNFFY – GYUCHA 2011). Hence, the situation which occurred in the summer of 2012, when archaeologists (to the developers' dismay) were forced to leave an unfinished and therefore untouchable excavation area after the set time limit was reached, could have been foreseen. On top of this, and besides the fact that the current law on cultural heritage protection in Hungary was designed without any input from professionals, the law also highly contradicts the international heritage protection conventions ratified by the Hungarian state.[8] The new measures taken by the government therefore poses a great threat to the Hungarian and European cultural heritage, which may come to suffer irreversible damages. Recently,[9] it does seem as if the government, perhaps because of a somewhat late realisation of their responsibility towards the national cultural heritage or due to the incitements of major developers, has taken certain steps towards improvement. For example, the State Secretary of Cultural Affairs recently hinted at the introduction of a new 1% tax to be paid by each developer (regardless of the existence of potential sites under the development area), which would then be deposited in a fund intended for large-scale salvage work. As it happens, this novel initiative matches our original proposal (see *Fig. 1*). So far the suggestion still stands and it can only be hoped for that nothing and nobody intervenes. That we are currently "walking on thin ice" is also reflected by a yet more recent decision of the Government,[10] concerning the demolition of the

---

7 In addition to Hungarian heritage protection organizations, some 20 European national and international associations, including EAA, AIA, SAA, ICOMOS, UISPP and WAC, also expressed their serious concerns about the law proposal in November 2011 emphasizing that it disharmonizes with European heritage protection traditions and practices as well as sharply contradicts the principles of the Valletta Treaty. On his facebook page Sir Colin Renfrew also indicated his worries.

8 These include the European Convention for the Protection of the Archaeological Heritage (Malta, 1992), the European Landscape Convention (Florence, 2000) and the Council of Europe Framework Convention on the Value of Cultural Heritage for Society (Faro, 2005). Ironically, the latter one has been ratified by the Hungarian Parliament in 2012.

9 On 30. August 2012.

10 On 6. Sept. 2012.

widely respected 130-year-old national institution, the Cultural Heritage Office (KÖH). No reasons for this action were either given, nor can such a step be justified in terms of heritage protection.

From the point of view of professional archaeologists we made the following suggestions to balance the historical interests of the public and the economic interests of developers: Although we certainly all strive to excavate every bit of heritage that is at risk of being destroyed, in regions like the Carpathian Basin it is unlikely to suppose any greater area to be devoid of sites or landscapes of historic importance. It becomes necessary to start learning how to gain the maximum information by a minimum size and amount of preventive excavations, but we also have to learn about a certain selection. Regarding this, the importance of the diagnostic methods used prior to excavation must be acknowledged, so that a very conscious and careful planning phase can provide the base for well informed decisions on what exactly to excavate, decisions which should also meet with strictly defined ethical codes.

The specific development of theoretical strands before and after "the change"

Mainly due to their history and culture, but also to the traditions within general education and university training, Hungarian archaeology has always been a part of the German school, that is to say the "typology-chronology" tradition that mostly consisted of mere archaeography and the description and typo-chronological classification of finds. This characteristic, partly persistent until today, was combined with historicism. In other words, archaeology used to be interpreted as a certain "*ancilla historiae*". Since many archaeologists still consider new strands and methods – as contrasted to "proper" archaeology – to be much of a cheat, the cultural historical paradigm is alive and well in Hungary. Moreover, they tend to think that those who practice these new ways will end up with a lot of money, but hardly any major sites, nor will they have a huge amount of finds. Thus, the conservative rejection of new ideas are often bound together with a certain false sense of pride, based on the idea that Hungarian archaeology has no need for such dubious methods, since it already has abundant and rich sites to dig and proceed.

On the whole, there are very few works that have discussed the state of theoretical archaeology in Hungary (Laszlovszki – Siklódi 1991; Chapman 2000; Bánffy 2004b; 2013; Jankovich-Bésán – Nagy 2004; Bartosiewicz *et al.* 2011). As all these works note that interestingly, but not standing without parallels in neighbouring countries, prehistory and especially Neolithic archaeology was the first to break out of the traditional frames.

It started in the 80s with the excavation of Neolithic tell sites. They were thought to be centres of domestication, hence the interest of the first zooarchaeologist, S. Bökönyi, who focused mainly on this period. When he later became director of the Institute of Archaeology HAS, his experience in environmental archaeology triggered him to establish two microregional research programs in Hungary. These projects, albeit they ended in useful publications in some respects (BÖKÖNYI 1996; VADAY *et al.* 2011), failed to fulfil the original intention of involving Hungary in the archaeological strands dealing with features and sites on a landscape-scale. This shortcoming was partly due to technical difficulties, e.g. that small-scale maps were of a military secret nature, restricted or forbidden to use, and that aerial photography was practically impossible for the very same reason. There were some other attempts to introduce environmental archaeology, but there was not a single site at which archaeological results would have been published *together* with scientific analyses. Lack of interest and misunderstandings in terms of "either/or" instead of "both", aggravated the introduction of new methods. It is only thanks to a group of devoted and bright personalities (e.g. the late Ede Hertelendi, or László Bartosiewicz and Pál Sümegi), that some development still happened, most notably in radiocarbon dating methods and in the fields of zooarchaeology, macrobotany, palynology or archaeogeological measurements. Still, the analyses were never constituted in a system for reconstructing palaeoenvironments, and they hardly had any impact on the discourses within archaeological research.

One might say this isolation or theoretical "backwardness" actually brought with it some benefits. We were left out of the sharp debate and exaggerated views that marked the formation of the new theoretical strands and by the time the gates opened, processual archaeology had already left its child diseases behind.

In the last eight years, the first outcomes of research incorporating both scientific methods and archaeological analyses have been published (SÜMEGI – GULYÁS 2004; ZATYKÓ *et al.* 2007; BÁNFFY 2004a; WHITTLE 2007; KREITER *et al.* 2012). Today, what can perhaps be described as the sunniest side of the changes that took place in the last two decades, are connected with processual archaeology. Besides newly initiated micro-regional projects (e.g. BÁNFFY 2006; PARKINSON *et al.* 2004), Bronze Age research has also finally, after decades of backlogging (mainly due to disinterest or personal politics) started up again with projects like the SAX project in Százhalombatta, the Benta Valley Program and surveys in the Körös Valley (VICZE – POROSZLAI 2000; EARLE – KRISTIANSEN 2010; DUFFY 2010). But these are only a few out of many recent developments; the now legal and wide-spread use of aerial photography has facilitated landscape-scale research and a GIS-lab has been founded in the ELTE Budapest University, there is now a geoarchaeological department in the Szeged University, a long-term cooperation has been established with the Geochemical Institute HAS, and a more and more widely known aDNA lab has been set up in the Arch. Inst. HAS. These are all hallmarks of a very important process: the involvement of scientific methods in archaeological research as part of general practice. In sharp contrast with earlier fieldworks, archaeologists are now working together with all kinds of experts on a regular basis in order to create detailed environmental, demographic, palaeodietary (etc.) reconstructions. It is now harvesting times, and all the work that has been done, all the information that has accumulated in this field since the wall fell, is put to use.

Postprocessual strands seem to have faced a much less happy fate in Hungary, perhaps because of the assumptions or impression shared by many, of these strands as being too psychologically involved, not compatible with a more factual way of thinking. As Higgs and Jarman (HIGGS – JARMAN 1975: 1) put it: "the soul leaves no skeleton behind". This statement does not only have bearing on some of the "isms", like neo-Marxist archaeology, which (understandably enough), enjoys no discursive space in the former socialist bloc whatsoever, but it also affects the potential impetus of social archaeology. Apart from a few colleagues abroad (CHAPMAN 1994; 1997; 2000; 2001; 2008; SOFAER DEREVENSKI 1997; SØRENSEN – REBAY-SALISBURY 2008), no one has introduced the narrative of social archaeology into the archaeological discourses in Hungary. Out of those few investigations that had been published earlier (for example a contextual analysis of cult objects within neolithic houses, published in 1991 by E. BÁNFFY [1990–1991]), none had managed to achieve the certain type of awareness or consciousness associated with what is called interpretative archaeology – to put it simply, nobody in Hungary had even heard of those works at that time. Similarly, the fallback is still great when it comes to areas like landscape or gender archaeology, which are barely mentioned in university courses. There are only a handful of professors who even try to change this situation. The only thing we can do at the moment is to offer such themes for PhD students, and by tutoring them individually we can help a new generation of "elite" archaeologists to learn, use and be aware of these approaches. In this way, embedded into the every day narratives of researchers, social or interpretative archaeology will last longer. Perhaps this delay will bring with it the same benefits as the late adoption of processual archaeology did, i.e. thanks to the delay, more settled ideas will reach us?

The dangers of distorted ideas in archaeology – an escalating social and political threat

A rather frightening phenomenon that has appeared after the change is the rise of fringe archaeological narratives. They are now accepted as real alternatives to professional research. This can be called a postmodern issue only in one sense, referring to its permissive, multiperspective starting point, in the sense that some postmodern scholars would agree to consider fringe and scientific theories as equally valid alternatives (SHERMER 1997: 24–61; FEDER 1990: 327). All the rest has nothing to do with postmodernity, nor actually with archaeology. Rather, it has to do with growing psychological, social and political unrest in contemporary Hungarian public life.

The discourse that flourishes in practically each of the post-socialist countries is very nationalistic in nature, and in the following section I shall try to give some information about its possible roots, the reasons for its quick spread and if there are any good strategies available for professional archaeologists to tackle this situation.

Before the change, there was practically no trespass between professional, academic archaeology and other interpretations of the past. This situation definitely carried a taste of authoritarian attitude, of 'academics' versus laymen, the upshot being that knowledgeable professionals enjoyed a certain amount of respect. It also meant that theories like the Sumerian origin of Hungarians were almost exclusively kept alive by immigrants who had left Hungary after World War II or in 1956.

After the change, suddenly, many taken for granted views became questioned and disputed. There were definitely some positive sides to this, but instead of a breeze of fresh air, more and more unscientific ideas surfaced along with with the enthusiasm of the newly gained political freedom. Support soon arrived from some of the former immigrants mentioned earlier. They came back to Hungary and offered to assist local "historians" by providing books and lectures that denounced official "lies" about Hungarian origin, "lies" like the Finno-Ugrian roots, which they claim was invented by the Habsburg Empire, and later endorsed by the Soviet

Empire, in order to humiliate and oppress the nation. There is no clear answer to the question of why alternative origins would somehow be more desirable, but apparently these easy-to-digest explanations are favoured over empirically based and compound historical, linguistic and archaeological interpretations. Also, the idea of conspiracy is important here, an ingredient that when mixed together with beliefs and theories like witchcraft and esotericism, Sumerian or Scythian-Hunnian-Etruscan origins, or with the idea that the Hungarian language and nation has existed for several millennia (some even say since the Palaeolithic), can become a very potent and dangerous formula. Supporters of these ideas claim that both scientific and unscientific statements are equally plausible and make for equally valid narratives. Or in reality, only the latter ones are considered valid, since 'academic' professionals are accused of betraying our homeland.

Contrasted to the Western New Age or fringe archaeological activity, alternative views have gained an added social and political significance in this part of Europe. These movements cannot be regarded as offshoots of other western groups of innocent and harmless "lunatics" as they have sometimes been called (e.g. FAGAN 1987), because each folkloristic and shamanistic festival or Ancient Hungarian pagan feast, comes with a powerful political message. This message is often symbolised by some "historic facts" like the Hungarian Runic script, which was used in Medieval Transylvania and which is now "detected" almost everywhere in European prehistory. Moreover, modern street- or village signs now tend to appear both in Latin and "Hungarian Runic" script (*Fig. 2*).

These Hungarian signs seem to appear in neolithic representations as well. One excavator and publisher who was always reluctant to read the signs in Old Hungarian, is now claimed to be "Moscovite" (*Fig. 3*).[11] Moreover, Hungarian script is apparently no longer restricted to the Carpathian Basin. Runic signs found at the ill-famed site of the Bosnian "Pyramids" now reveal to us who the real builders were tens of thousands of years ago: Hungarians of course.[12]

Since these beliefs are often competing with the teachings of Christianity, the church has finally woken up just recently, only to find that it is already too late to raise its word against this new paganism. Certainly, the dangers lies not in these sagas themselves, but in the fact that they are all connected with the denial of

---

[11] http://forum.index.hu/Article/jumpTree?a=78358068&t=9119978 [21.04.2008]; http://www.szentgyorgyvolgy.hu/banffy-eszter-es-szentgyorgyvolgyi-tehenszobor-szojelei; http://www.nemzetihang.hu/modules.php?name=Content&pa=showpage&pid=19 [10.09.2008].

[12] http://www.youtube.com/watch?v=7HwzImhmpkU [20.07.2012].

a

b

Fig. 2. Places in the Carpathian Basin signposted by Hungarian runic scripts.
a. Beregovo-Beregszász, Ukraine (wikipedia.org – cc by-sa.3.0).
b. Zugló, district XIV. in Budapest, with signs of passing judgement of local inhabitants (photo: author 2012)

a

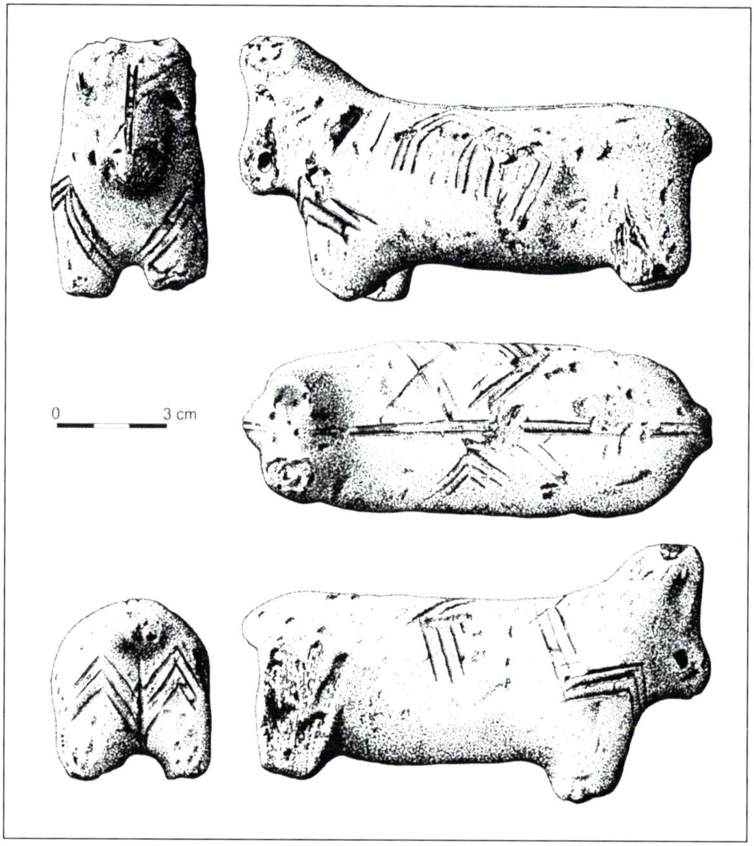

b

Fig. 3. "Hungarian" runic signs on a 6th Millennium BC figurine from West Hungary. a. Photo (after BÁNFFY *2004a). b. Drawing (after* BÁNFFY *2004a)*

the borders established in the Trianon Treaty, and are eagerly claiming back the lost two thirds of Hungary's territory before 1920. This Old, Greater Hungary, appears in various pictographic forms as part of their sociopolitical message, together with narratives of our glorious past.[13]

Such narratives always occur together with EU-scepticism, antisemitism and a deep seated hatred for Roma people, homosexuals and other minorities. Radical groups calling themselves the "real" or "deep" Hungarians (while they use the term "tenuous" Hungarians for democrats), play a crucial role in street fights and violent assaults on the former, just as it happened to supporters of the left-wing government some years ago, or as it happens today against Roma village communities. The ultra-right demonstrations are always swamped with flags and garments resembling simultaneously ancient "Hungarian" symbols like the runic script, imitations of archaeological finds from the 10th century Hungarian Conquest period, and types of garments that were in use in between the two world wars and during the Nazi government in 1944.

From this observation follows, that the groups which can be called neo-Nazis use all these ideologies and symbols as their own. And before anyone start to think that this is just some type of fringe phenomenon, a tiny and insignificant ultra-nationalistic group, take a look at the outcomes of the last elections. They speak for themselves: the far right JOBBIK party has got more than ten percent of the votes and they sit in the European parliament.

According to the plans of the JOBBIK party, a new institute for Hungarian Prehistory is to be established, promoting research which would help replace the Finno-Ugrian roots with more glorious accounts of the Hungarian past. These "improved" accounts and theories would then not only be made available to schools, but become an obligatory part of the general education system. The ruling right wing government silently complies with their ideology. These days the process is already at a stage where it is recommended for schools to teach alternative history. The minister of education recently explained that children should get familiar with other histories than just the official version (or what has been considered as such up until now), so that they – the children(!) can judge.

Yet, as I wish to concentrate on possible social explanations rather than on politics, I will now try to provide some ideas that may lie behind this frightening phenomenon.

---

[13] Just one example among many: http://www.parameter.sk/blog/csehszlovak-kem/2013/06/17/igy-gondozzuk-trianon-traumankat [17.06.2013].

First, we should mention the *loss of economic stability*. After the change, a large proportion of post-socialist society firmly believed that immediate wealth would accompany the introduction of capitalism, and that their lives would be guided by the paternalistic state just like before: it was far easier to expect that the rapid climb to Western European standards was a matter for the state and not something they had to achieve on their own. Needless to say, instead of these expectations being fulfilled, along with the bankruptcy of socialist construction and heavy industry, millions lost their jobs and ended up in a hopeless situation. Fantastic ideas spread rapidly among these groups, as a sort of projection of their self dignity and a search for compensation.

Besides repeated economic crises, a gradual *loss of culture* is an escalating problem. Beginning this year, schoolchildren are not expected or even encouraged to acquire any firm knowledge of history, and according to the new education law the length of the obligatory education is to be decreased (from 18 to 15/16, not yet decided). Based on this scenario, if they hear about pseudohistorical theories, how will they be able to form any opinion of their own? For uncultured young adults, a well-equipped low-level world is offered, a world which consists of soap operas, commercial TV channels and the simple, quitchy music, offered by "national rock" groups (like the Kárpátia rock band). The texts of their songs are utterly nationalistic and often racist.

Losing social stability can be regarded partly as a consequence of the first two issues. In Hungary, during the last hundred years, there has has been no less than eleven changes of political systems (including the turn since 2010). "Up" groups have become "down" groups, always feeding the political and cultural distinctions of "us" versus "them". This has led to uncertainty and revived the idea of loneliness, of the feeling that "everyone is against us". An emotionally unstable mind leads to an equally unstable identity. In order to help such a damaged ego, the world is often sharply divided into black and white: of perfectly good people and truly wicked people, groups or ideas. Whatever comes from the good side will be uncritically accepted, whilst everything from the bad side will be rejected. This split identity does not allow the person to accept anything, even the most plausible facts and thoughts uttered by "enemies", because this would be the same thing as acknowledging their ultimate superiority, a recognition which would cause further damages to their identity.[14] Seen in this way, their hostile

[14] I wish to thank the psychologist Eszter Fischer for sharing her ideas on this issue with me.

attitude towards the teachings and representatives of the realm of science can be understood and explained. Everything becomes suspicious if said by 'academics'. Professional archaeologists have landed in the bad book.

This distressed and unsettled type of identity, when bolstered by an uncultured background, "neopaganism", alternative world theories and some distorted historic origin myths, will be an easy prey for manipulations. Hungary, together with other former socialist countries has not yet dealt with the dark parts of it past, the "*Vergangenheitsbewältigung*". To blame the world for two lost world wars and a four-decade-long Soviet occupation has led to cognitive dissonance, a state in which conspiracy theories and ideas of treachery often thrives. According to these, Hungary was oppressed *in spite* of being a great nation – Hungary was oppressed *because* of being a great nation. As already mentioned, one such idea is that the Finno-Ugrian origin was invented by the Habsburgs in order to humiliate Hungarians and to destroy their self dignity. Later on, this "cruel invention" was ascribed to the Soviets, and now to the Jews and the globalised capitalist system. But why is it considered so shameful to be related with the Finns? There is no good answer, unless we consider the romanticised pictures of steppic people on horseback armed to their toes and killing all enemies, superior to those of people fishing peacefully at the waterfront. All things considered: the emergence of unstable and damaged identities in Hungarian society, combined with an eagerness for compensation, a susceptibility to grand and emotionally charged narratives, and a declining educational system promoting erroneous and uncritical thought – one can safely say that this is a situation that nobody had envisioned at the fall of the Berlin wall, there on the threshold to liberty. As demonstrated in the previous discussion about what potential reasons may lay behind the current state of affairs in the country, sociological analyses has showed that there is a strengthening intolerance against certain groups of people, xenophobia and hatred is gaining more and more ground (see ÖRKÉNY – VÁRADI 2010). In the midst of uncertain social and psychological circumstances, distorted historical theories function like an Ariadne's thread in the labyrinth: they provide a feeling of security. Since Hungary has always been – and still is – at the crossroads of north-south and east-west migration and important cultural movements, the dangers I have shortly discussed in this paper can hardly be considered as limited, or as an isolated symptom.

There are few possibilities for professional archaeologists to improve this state, or to change the negative image of archaeology broadcasted by the media and nationalist movements. In order to balance distorted theories with facts and to show how impressive they could be, we need to publish our results in all sorts of media, especially on the internet. Interestingly enough, these distorted archaeologies often spread extremely fast and circulate intensely on hundreds of sites and chat-forums. In a Google search on the word "prehistory", the first dozen hits in the result list link to unscientific sites, refreshed on a daily basis. Blogs, newsletters, virtual museum videos and more excavation reports disseminated by professionals, are all essentials if archaeologists are to stand a chance of making their voices heard. Another difficult task is to maintain the dialogue with these groups and reply to fringe questions, at least up to the point at which there is still a chance for an understanding reception. But we must always bear in mind that, since this phenomenon is not restricted to archaeology, it cannot be solved on its own, and certainly not by the relatively small group of professional archaeologists in Hungary. Any turn towards an improvement of the current situation necessitates a profound shift in social cognition as well. In other words it is a change that can hardly come about without the improvement of hands-on social, cultural, educational circumstances. And that will be a lengthy process.

Concluding thoughts

In the beginning, I said I would pick up three themes among the multitude of changes that has occurred in Hungarian archaeology and cultural heritage since 1989. Apart from the fact that they all emerged at the same time and had a definite effect on attitudes to the past, they also seem to be related to each other in other ways, due to different aspects in their common background. More precisely, this can be called the benefit and drawback of freedom: broad possibilities on the one hand, but without any major dividing scientific disputes, and eventually, on the other we face the exploitation of this anarchy.

In socialist times the state ruled over institutions, programs and thoughts. When this strong control weakened within a short period of time, abating in only one or two years, both professional archaeologists and other claimants to the past suddenly felt the taste of unbound freedom. Professionals started to read what had earlier been inaccessible literature and to use previously unthinkable scientific methods. This led to new contacts being established and triggered plenty of new cooperation with good results, but the amount of trajectories, theories and methods also caused certain confusion. However, the overwhelming majority of professional archaeologists never left their field of cognition and would not be influenced by emotions, let alone any kind of political wish.

A group of non-professionals moved the opposite way. The unscientific or distorted ideas discussed above, became interpreted as ideas that were forbidden in socialist times. Consequently, they began a new career as "the hidden truth", previously withheld by "communists". This alone makes such strands (reclaiming a glorious Hungarian past, after a thousand-year-long suppression of the nation by alien forces etc.) quite different from romantic fringe archaeologies in Western Europe or North America. In a Hungarian context, professional archaeologists who stick with dry facts and scientific methods are often judged harshly by a rather hostile jury. This judgement comes not only from harmless "lunatics", but also from "deep" Hungarian groups on the far right. It was when when these attitudes recently became part of national politics that things started to become really dangerous: views now powerfully expressed and enforced by the far-right JOBBIK party in the Parliament, and at a growing rate also in official discourse.

The third point I picked up may seem, at first glance, as an entirely coincidental phenomenon, perhaps with a far-fetched link to the new possibilities opening up after the change, opportunities to construct motorways and similar large-scale projects. Nevertheless, since large-surface preventive archaeology is expensive it has clearly become an obstacle for both developers and the government, leading to new or reinterpreted laws, offering little or no support for neither professional archaeologists nor the cultural heritage of Hungary. If the previously strong position of professionals had not been made relative by politicians, the media and various hostile groups questioning the validity of their work, the damages on the archaeological heritage (mentioned in the text and in the footnotes) could not have happened.

By drawing a sketchy picture of some of the characteristics visible in one post-socialist country, Hungary, I certainly cannot, and do not want to make any general inferences, either on the state of East European heritage and archaeology, or on their impact on social or political narratives. Still, while talking to colleagues from other post-socialist countries I have experienced similarities with several of the points touched upon in this paper. In the TAG meeting in Durham 2009, there was a session organised by Bisserka Gaydarska and John Chapman, focusing on archaeology in former socialist countries and the changes of the past twenty years, after the fall of the Berlin wall. Some of the papers presented there, especially the ones dealing with developments in the former Yugoslavia and Ukraine, revealed rather similar problems. As I am convinced, it is plain talk that upholds rational thought.

References

Bánffy, Eszter (1990–1991), Cult and archaeological context in Central and South East Europe in the Neolithic and Chalcolithic. *Antaeus* 19–20, 183–250.

Bánffy, Eszter (2004a), *The 6th millennium BC boundary in Western Transdanubia and its role in the Central European transition (The Szentgyörgyvölgy-Pityerdomb settlement).* Varia Archaeologica Hungarica 15. Budapest.

Bánffy, Eszter (2004b), *What is time? A secret – On the present state of archaeological research in Hungary.* Hungarian Academy of Sciences 2004.11, 1240–1246. (http://www.matud.iif.hu/04nov/08.html [17.07.2013]).

Bánffy, Eszter (ed.) (2006), Archaeology and settlement history in the Kerka valley – SW Hungary. *Antaeus 28.*

Bánffy, Eszter (2013), Theoria est ancilla rationis politicae? The "Neolithic revolution" in the archaeological research of socialist Hungary. In: Sophie Bergerbrant – Serena Sabatini (eds), *Counterpoint: essays in archaeology and heritage. Studies in honour of Kristian Kristiansen.* British Archaeological Reports International Series 2508. Oxford. 623–630.

Bánffy, Eszter – Attila Gyucha (2011), Fundamental modifications to archaeological heritage protection regulations in Hungary: a brief report. *The European Archaeologist* 36, 5–6.

Bánffy, Eszter – Pál Raczky (2010), The crisis and changes in the cultural heritage legislation in Hungary: cul-de-sac or solution? In: Nathan Schlanger – Kenneth Aitchison (eds), *Archaeology and the global economic crisis: Multiple impacts, possible solutions.* ACE – Culture Lab Editions. Tervuren. 81–86.

Bartosiewicz, László – Dóra Mérai – Péter Csippán (2011), Dig up – Dig in: practice and theory in Hungarian archaeology. In: Ludomir R. Lozny (ed.) *Comparative archaeologies: A sociological view of the science of the past.* New York. 273–337.

Behrens, Hermann (1984), *Die Ur- und Frühgeschichtswissenschaft in der DDR von 1945–1980: miterlebte und mitverantwortete Forschungsgeschichte.* Frankfurt am Main – New York.

BIEHL, PETER F. – ALEXANDER GRAMSCH – ARKADIUSZ MARCINIAK (2002), Archaeologies of Europe: histories and identities. An introduction. In: Peter F. Biehl – Alexander Gramsch – Arkadiusz Marciniak (eds), *Archaeologies of Europe. History, methods and theories / Archäologien Europas. Geschichte, Methoden und Theorien.* Tübinger Archäologische Taschenbücher 4. Münster – New York – München – Berlin. 25–31.

BÖKÖNYI, SÁNDOR (ed.) (1996), *Cultural and landscape changes in South-East Hungary II: prehistoric, Roman barbarian, and Late Avar settlement at Gyoma 133 (Békés County Microregion).* Archaeolingua Main Series 5. Budapest.

CHAPMAN, JOHN C. (1994), Social power in the early farming communities of Eastern Hungary – perspectives from the Upper Tisza region. *Jósa András Múzeum Évkönyve* 36, 79–99.

CHAPMAN, JOHN C. (1997), The origin of tells in Eastern Hungary. In: Peter Topping (ed.), *Neolithic landscapes.* Neolithic Studies Groups Seminar papers 2. Oxbow Monographs 86. Oxford. 139–164.

CHAPMAN, JOHN C. (2000), Research issues in Hungarian prehistory. Introduction to: *Tensions at funerals: mortuary archaeology in later Hungarian prehistory.* Archaeolingua Series Minor 14. Budapest.

CHAPMAN, JOHN C. (2001), The fractality of personal relations in the Mesolithic and Neolithic South East Europe. In: Róbert Kertész – János Makkay (eds), *From the Mesolithic to the Neolithic. Proceedings of the International Archaeological Conference held in the Damjanich Museum of Szolnok, Sept. 22–27, 1996.* Budapest. 145–165.

CHAPMAN, JOHN C. (2008), Meet the ancestors: settlement histories in the Neolithic. In: Douglass Whitfield Bailey – Alasdair Whittle – Daniela Hofmann (eds), *Living well together? Materiality in the Neolithic of South-East and Central Europe.* Oxford. 68–80.

DUFFY, PAUL R. (2010), *Complexity and autonomy in Bronze Age Europe: assessing cultural developments in the Eastern Hungary.* Unpublished Ph.D. thesis, University of Michigan, Ann Arbor.

EARLE, TIMOTHY K – KRISTIAN KRISTIANSEN (eds) (2010), *Organizing Bronze Age societies: the Mediterranean, Central Europe, and Scandanavia compared.* Cambridge.

FAGAN, BRIAN (1987), Archaeology and pseudo-archaeology. In: *University of Pennsylvania Museum of Archaeology and Anthropology. Documents* 29.2, 2–3. (http://www.penn.museum/documents/publications/expedition/PDFs/29-2/Archaeology.pdf [10.07.2013]).

FEDER, KENNETH (1990), *Frauds, myths, and mysteries: science and pseudoscience in archaeology.* Mountain View.

HIGGS, ERIC S. – MICHAEL R. JARMAN (1975), Palaeoeconomy. In: Eric S. Higgs – Michael R. Jarman (eds), *Palaeoeconomy*. Cambridge. 1–7.

JANKOVICH-BÉSÁN, DÉNES – MIHÁLY NAGY (2004), *Felmérés a régészet helyzetéről 1989–1999.* [Survey on the state of archaeology 1989–1999] A Nemzeti Kulturális Örökség Minisztériumának Műemléki Főosztálya [Hungarian Ministry of Cultural Heritage, Department for Monuments and Sites]. Budapest.

KAISER, TIMOTHY (1995), Archaeology and ideology in southeastern Europe. In: Philip L. Kohl – Clare Fawcett (eds), *Nationalism, politics and the practice of archaeology*. Cambridge. 99–119.

KREITER, ATTILA – ÁKOS PETŐ – BEÁTA MAGYAR TUGYA (eds) (2012), *Környezet – Ember – Kultúra. A természettudományok és a régészet párbeszéde Magyar Nemzeti Múzeum Nemzeti Örökségvédelmi Központ 2010. Október 6–8-án megrendezett konferenciájának tanulmánykötete / Environment – human – culture. Dialogue between applied sciences and archaeology. Proceedings of the conference held between 6th and 8th of October 2010 by the National Heritage Protection Centre of the Hungarian National Museum.* Budapest.

KUNA, MARTIN – NATALIE VENCLOVÁ (1995), Introduction. In: Martin Kuna – Natalie Venclová (eds), *Whither archaeology? Studies in honour of Evžen Neustupný.* Prague. 7–10.

LASZLOVSZKY, JÓZSEF – CSILLA SIKLÓDI (1991), Archaeological theory in Hungary since 1960: theories without theoretical archaeology. In: Ian Hodder (ed.), *Archaeological theory in Europe: The last three decades.* London – New York. 272–298.

MILISAUSKAS, SARUNAS (1986), Selective survey of archaeological research in Eastern Europe. *American Antiquity* 51, 779–798.

MILISAUSKAS, SARUNAS (1990), People's revolutions of 1989 and archaeology in Eastern Europe. *Antiquity* 64, 283–285.

Örkény, Antal Székelyi Mária – Luca Varádi (2010), Az előítéletes gondolkodás társadalmi beágyazottsága, nemzetközi összehasonlításban [Social embeddedness of prejudicial thinking, an International Comparison]. *Alkalmazott Pszichológia* 12.1/2, 29–46.

Parkinson, William A. – Richard W. Yerkes – Attila Gyucha (2004), The transition to the Copper Age on the Great Hungarian Plain: The Körös regional archaeological project. Excavations at Vésztő-Bikeri and Körösladány-Bikeri, Hungary, 2000–2002. *Journal of Field Archaeology* 29, 101–121.

Shermer, Michael (1997), *Why people believe weird things: Pseudoscience, superstition, and other confusions of our time*. New York.

Sofaer Derevenski, Joanna (1997), Age and gender at the site of Tiszapolgár-Basatanya, Hungary. *Antiquity* 71/274, 875–889.

Sørensen, Marie Louise Stig – Katharina Rebay-Salisbury (2008), Landscapes of the body: Burials of the Middle Bronze Age in Hungary. *European Journal of Archaeology 11.1*, 49–74.

Sümegi, Pál – Sándor Gulyás (eds) (2004), *The geohistory of Bátorliget Marshland. An example for the reconstruction of Late Quaternary environmental changes and past human impact from the northeastern part of the Carpathian Basin.* Archaeolingua Main Series 16. Budapest.

Vaday, Andrea H. – Dénes Jankovich-Bésán – László Kovács (2011), *Archaeological investigations in county Békés: 1986–1992.* Varia Archaeologica Hungarica 25. Budapest.

Vicze, Magdolna – Ildikó Poroszlai (2000), *SAX Százhalombatta Archaeological Expedition Annual Report 1 – Field Season*. Százhalombatta.

Whittle, Alasdair (ed.) (2007), *The early Neolithic on the Great Hungarian Plain. Investigations of the Körös culture site of Ecsegfalva 23, County Békés.* Varia Archaeologica Hungarica 21. Budapest.

Zatykó, Csilla – Imola Juhász. – Pál Sümegi (eds) (2007), *Environmental archaeology in Transdanubia.* Varia Archaeologica Hungarica 20. Budapest.

About the authors

Eszter Bánffy is an archaeologist specialising in the Neolithic and Chalcolithic of Central and South East Europe, currently paying special attention to the Central European Neolithic transition. She is also dealing with theoretical issues and matters of heritage protection. As a professor of prehistoric archaeology and geoarchaeology, she is in charge of a work team and has conducted several Hungarian projects as well as projects with wider international connections. She has authored seven books and more than a hundred chapters and articles, published in Europe and America. She has given lectures and courses in European and American Universities including Harvard, where she spent a semester as visiting scholar in 2008. She served as executive board member and secretary of the European Association of Archaeologists (EAA) (2005–2011). Besides working at the Institute of Archaeology of the Hungarian Academy of Sciences, she is supervising PhD students at two universities: ELTE Budapest and SZTE Szeged. Beginning in 2014, she will suspend the vast majority of these tasks as she will start to work as the director of the Roman-Germanic Commission of the German Archaeological Institute in Frankfurt.

John Bintliff (London, England 1949) is a Professorial Fellow at Edinburgh University, United Kingdom, and Professor of Classical and Mediterranean Archaeology at Leiden University, the Netherlands. He studied Archaeology and Anthropology at Cambridge University, where he also completed his PhD in 1977 on the (pre)history of human settlement in Greece. He was Senior Lecturer in Archaeology at Bradford University, where he taught from 1977, then moved to Durham University as Reader in Archaeology in 1990, where he taught till moving to Leiden in 1999. In 1988 he was elected a Fellow of the Society of Antiquaries. Since 1978 he has been co-directing (with Cambridge University) the Boeotia Project, an interdisciplinary programme investigating the evolution of settlement in Central Greece, widely-recognised as one of the most significant regional research programmes in the Mediterranean region. Since his first paper in 1975, he has published on average 4,5 academic publications per year, amongst which are 20 books.

Michael A. Cremo is an independent historian of archaeology. He lives in Los Angeles. He is the author of the books *Forbidden Archeology, Human Devolution:*

A Vedic Alternative to Darwin's Theory, and *My Science, My Religion,* a collection of twenty-four papers presented at academic conferences. He is a member of the European Association of Archaeologists, the World Archaeological Congress, and the European Association for South Asian Archaeology and Art. His work in the history of archaeology is influenced and inspired by his studies of the *Puranas,* the historical/cosmological texts of ancient India.

Gerhard Ermischer was born in Salzburg (Austria) in 1963 and studied history and archaeology in Innsbruck (Austria) and Southampton (UK). He wrote his doctoral thesis in medieval history in Rome. In 1991 he came to Aschaffenburg (Germany) in order to work on the material from city excavations, where he later became the city archaeologist and curator at the city museums. He founded the Archaeological Spessart-Project (ASP) in 1998 as a regional association for archaeology and research in the Spessart Region. It developed into a leading institute on landscape studies and communication, with a unique approach to participative research. Since 2007 he works solely as chair of the institute, which became an Institute at the University of Würzburg in 2010 and was awarded the Agenda-21 Award of the City of Aschaffenburg in 2011 and the Bavarian Award for Archaeology in 2013.

Hamish Forbes recently retired as Associate Professor and Reader in Anthropological Archaeology in the Department of Archaeology at the University of Nottingham. His research interests primarily involve the integration of ethnographic and archaeological approaches in the Mediterranean region, particularly Greece. His main focus is on social issues, and environmental concerns relating to agriculture, pastoralism and the meanings of landscapes. Recent publications include *Meaning and Identity in a Greek Landscape: an Archaeological Ethnography* (CUP, 2007), and 'Lost soles: ethnographic observations on manuring practices in a Mediterranean community' (in: R. Jones (ed.), *Manure matters: historical, archaeological and ethnographic perspectives*; Ashgate, 2012).

Stella Katsarou (PhD) is affiliated with the Department of Palaeoanthropology-Spelaeology, Athens. She has been working in the academic faculty at the University of the Aegean, and guest lecturer at the 'Arizona in the Aegean Program' run by the University of Arizona. She specializes in Neolithic pottery and participates in several research projects at prehistoric sites, both on the

Greek mainland and in the Aegean. Projects include the Tell site of Koutroulou Magoula, carried out in collaboration with the University of Southampton, the Helike Project, and the caves of Theopetra and Diros in collaboration with the Institute for Aegean Prehistory and the American School of Classical Studies in Athens. Her academic focus also includes the historically situated perceptions of prehistory and its role in the present. Stella Katsarou has received grants from various institutions and is extensively published.

Tuija Kirkinen (born 1963) is a Finnish archaeologist working as an amanuensis at the department of archaeology at University of Helsinki. She is currently writing a dissertation that deals with the land-use history in the Eastern Finnish wilderness areas. She specializes on GIS and landscape archaeology and has published several studies on these topics.

Jutta Leskovar is a Prehistorian at the Upper Austrian State Museum. Her research interests involve Iron Age, Pagan, Heritage and Gender Studies. Her work is focused on the transition zone between archaeological and public interests of different kinds. She is responsible for the open-air museum of Mitterkirchen (site of a Hallstatt period cemetery), and is engaged regularly as lecturer at the University of Vienna.

Since 2008 **Thomas Meier** is professor for pre- and protohistory at the University of Heidelberg (Germany). Before that he held a visiting professorship in Jena. In 1999 he was rewarded a PhD at the University of Munich for a thesis on royal tombs in medieval Europe. From 1997 until 2006 he was heading a project on human-environmental interaction in the Middle Ages, focused on a small priory in the Bavarian Alps. In his habilitation at the University of Kiel, he worked on the social structure and economic system of Carolingian southern Germany. Actually, Thomas Meier researches and publishes mainly on archaeological theory, the spatial dimensions in archaeology, environmental and landscape archaeology.

In 2011 **Elisabeth Niklasson** started her PhD studies at the Department of Archaeology and Classical Studies at Stockholm University. As part of the interdisciplinary research school Fokult, she pursues her interest in the socio-political aspects of archaeology and in cultural heritage formation as phenomenon and practice. Combining texts and ethnographically collected sources, her PhD project explores the practices, driving forces and conditions underlining the interaction between archaeologists and the European Commission, especially as seen through funding programmes. Before this she participated in archaeological research projects in Tanum and on Gotland in Sweden, as well as in Bosnia, Ireland and Argentina, especially dealing with GIS and digital documentation.

The archaeologist **Diane Scherzler** is part of the management team of SWR, a German public broadcasting corporation. From 2002 to 2012 she worked as a science journalist at SWR. For almost 20 years Diane Scherzler has worked on the complex of themes "science, media and public". She teaches science communication at several universities, and for many years she has held media training courses for scientists who want to improve their collaboration with journalists. She is also a Vice Chairperson of the German Society for Primeval and Early History (DGUF) and publishes the monthly DGUF newsletter. Diane Scherzler founded the Euroscience work group "Science Communication" and also heads it. She is also one of the co-founders of the "Kulturgut-Raub" (looting of cultural assets) Facebook group which critically examines the global illegal trade in antique artefacts and the destruction of cultural assets. The group has almost 250 members.

ARCHAEOLINGUA

Edited by
ERZSÉBET JEREM and WOLFGANG MEID

Main Series

24. **The Master of Animals in Old World Iconography.** Edited by Derek B. Counts and Bettina Arnold. 2010. 256 pp. € 56.-. ISBN 978-963-9911-14-7.

25. **Archaeological, Cultural and Linguistic Heritage. Festschrift for Erzsébet Jerem in Honour of her 70th Birthday.** Edited by Peter Anreiter, Eszter Bánffy, László Bartosiewicz, Wolfgang Meid and Carola Metzner-Nebelsick. 2012. 640pp. € 88.-. ISBN 978-963-9911-28-4.

26. **The Levantine Question / El problema "Levantino".** (bilingual) Edited by José Julio García Arranz, Hipólito Collado Giraldo and George Nash. 2012. 432 pp. € 74.-. ISBN 978-963-9911-31-4.

27. **The Medieval Royal Palace at Visegrád.** Edited by Gergely Buzás and József Laszlovszky. 2013. 398 pp. € 72.-. ISBN 978-963-9911-39-0.

28. Anthony Harding and Valerii Kavruk: **Explorations in Salt Archaeology in the Carpathian Zone.** 2013. 332 pp. € 66.-. ISBN 978-963-9911-44-4.

29. **Textiles from Hallstatt. Weaving Culture in Bronze Age and Iron Age Salt Mines / Textilien aus Hallstatt. Gewebte Kultur aus dem bronze- und eisenzeitlichen Salzbergwerk.** (bilingual) Edited by Karina Grömer, Anton Kern, Hans Reschreiter and Helga Rösel-Mautendorfer. 2013. 572 pp. € 78.-. ISBN 978-963-9911-46-8.

Series Minor

28. Sofia Pescarin: **Reconstructing Ancient Landscape**. 2009. 264 pp. € 38.-. ISBN 978-963-9911-09-3.

29. **Situating Gender in European Archaeologies.** Edited by Liv Helga Dommasnes, Tove Hjørungdal, Sandra Montón-Subías, Margarita Sánchez Romero and Nancy L. Wicker. 2010. 309 pp. € 36.-. ISBN 978-963-9911-15-4.

30. **A History of Central European Archaeology.** Edited by Alexander Gramsch and Ulrike Sommer. 2011. 219 pp. € 24.-. ISBN 978-963-9911-23-9.

31. **Archaeological Imaginations of Religion.** Edited by Thomas Meier and Petra Tillessen. ISBN 978-963-9911-24-6. FORTHCOMING

32. Mária Bondár: **Prehistoric Wagon Models in the Carpathian Basin.** 2012. 142 pp. € 28.-. ISBN 978-963-9911-34-5.
33. **Appropriate narratives. Archaeologists, publics and stories.** Edited by Elisabeth Niklasson and Thomas Meier. 2013. 298 pp. € 34.-. ISBN 978-963-9911-47-5.

EAC series

1. EAC Occasional Paper No. 1. **The Heritage Management of Wetlands in Europe.** Edited by Byrony Coles and Adrian Olivier. 2001. 207 pp. € 15.-. ISBN 90-76972-01-X.
2. EAC Occasional Paper No. 2. **Europe's Cultural Landscape: archaeologists and the management of change.** Edited by Graham Fairclough and Stephen Rippon. 2002. 234 pp. € 22.-. ISBN 90-76975-02-X.
3. EAC Occasional Paper No. 3. **Listing Archaeological Sites, Protecting the Historical Landscape.** Edited by Peter A. C. Schut. 2009. 169 pp. € 38.-. ISBN 978-90-579-9144-8.
4. EAC Occasional Paper No. 4. **Heritage Management of Farmed and Forested Landscapes in Europe.** Edited by Stephen Trow, Vincent Holyoak and Emmet Byrnes. 2010. 184 pp. € 30.-. ISBN 978-963-9911-17-8.
5. EAC Occasional Paper No. 5. **Remote Sensing for Archaeological Heritage Management.** Edited by David C. Cowley. 2011. 312 pp. € 40.-. ISBN 978-963-9911-20-8.
6. EAC Occasional Paper No. 6. **Large Scale Excavations in Europe. Fieldwork Strategies and Scientific Outcome.** 2012. Edited by Jörg Bofinger and Dirk Krausse. 208 pp. € 35.-. ISBN 978-963-9911-29-1.
7. EAC Occasional Paper No. 7. **Heritage Reinvents Europe.** 2013. Edited by Dirk Callebaut, Jan Mařík and Jana Maříková-Kubková. 224 pp. € 36.-. ISBN 978-963-9911-41-3.
8. EAC Occasional Paper No. 8. **Who cares? Perspectives on Public Awareness, Participation and Protection in Archaeological Heritage Management.** 2013. Edited by Agneta Lagerlöf. 136 pp. € 28.-. ISBN 978-963-9911-42-0.

Please address orders to:

ARCHAEOLINGUA

H-1250 Budapest, Pf. 41.

Fax: (+361) 3758939

e-mail: kovacsr@archaeolingua.hu

http://www.archaeolingua.hu www.hungarianarchaeology.hu